The Power of Science

Economic research and European decision-making: The case of energy and environment policies

P.I.E. Peter Lang

Bruxelles · Bern · Berlin · Frankfurt am Main · New York · Oxford · Wien

Domenico Rossetti di Valdalbero

The Power of Science

Economic research and European decision-making: The case of energy and environment policies

The opinions expressed in this book are the sole responsibility of its author and they do not necessarily reflect the ones from the European Commission.

1 avenue Maurice, B-1050 Brussels, Belgium
info@peterlang.com; www.peterlang.com

ISBN 978-90-5201-586-6
D/20010/5678/05

Library of Congress Cataloging-in-Publication Data
Rossetti di Valdalbero, Domenico.
The power of science: economic research and European decision-making: the case of energy and environment policies
/ Domenico Rossetti di Valdalbero.
p. cm. Includes bibliographical references. ISBN 978-90-5201-586-6
1. Energy policy—European Union countries. 2. Power resources—Environmental aspects—European Union countries. I. Title.
HD9502.E92R67 2010 333.79094—dc22 2009044100

CIP also available from the British Library, GB.

Bibliographic information published by "Die Deutsche Nationalbibliothek"
"Die Deutsche Nationalbibliothek" lists this publication in the "Deutsche Nationalbibliografie"; detailed bibliographic data is available on Internet at <http://dnb.d-nb.de>.

Table of Contents

Foreword

Shaping the future has always interested me. I often had the opportunity to benefit from the most recent scientific results to inform my policy choices. At the European-level, this was the case during my tenure as Head of Cabinet of Jacques Delors, the President of the European Commission between 1984 and 1994. *The 1992 challenge from Europe – Development of the European Community internal market*, also known as the research on the *Cost of non-Europe* is a clear example of this. This "Cecchini Report" was largely based on different quantitative and qualitative methods that are well described by Domenico Rossetti di Valdalbero in the initial part of his book.

The quantitative approach (modelling) has the advantage of providing a systemic view to gain a better understanding of a clear techno-economic problem. The qualitative approach includes expert opinions and Delphi surveys, and allows the research community to put issues into a wider socio-political context. Both approaches are as necessary as they are complementary. Choosing the right approach is essential to gain real insights into the potential of policy to bring about real and transformative change. These two approaches were used – and are currently widely used – to support European policy-making.

While citizens should be conscious of the past to forge their own identity, policy-makers should anticipate future challenges. This is a prerequisite if we are to face uncertainty and try to take the best possible decision in the present. It was precisely this idea that led to the creation the European Commission Forward Studies Unit. It was also this principle which inspired the so-called European financial perspectives (initially called "Delors Package"), the long-term multi-annual visions of European spending needs, which are the financial means for the EU's future hopes and dreams.

At times this book reads like a journey down memory lane. It includes a literature review beginning with the launching of the Research "Framework Programme" in the middle of the 1980s. It goes on to explain how through successive Research Framework Programmes, scientists were encouraged to see how transformative change in the field of research might best be made into working solutions for the people of Europe.

Over the last twenty-five years, academics have worked at analysing the impact of European economic and social cohesion, monetary and enlargement policies. Since the beginning of the 21st century, questions around globalisation and a knowledge-based society (including human capital, the use of information technologies and the role of innovation) have been at the core of the European political and research agenda.

Domenico Rossetti's book allows for a thoughtful reflection on the recent economy-energy-environment paradigm change. It is an issue that is directly related to the quality of life of citizens of the world and the preservation of important global resources for future generations. Moreover, this paradigm speaks to the two major challenges facing both the developed and the developing world – security of energy supplies and sustainable development.

The European policy case studies addressed in this book deal with energy and environmental targets (including the increase share of renewables, the improvements in energy efficiency and the reduction of greenhouse gas emissions), with market instruments like taxation and emission trading, and supported by the principle of the "internalisation of external costs". It highlights a large part of the scientific debate about analysing the future (foresight and forward-looking) and the major scientific controversies surrounding economic and energy modelling.

This book comes at a critical juncture: severe fluctuations in oil prices, tensions around current and future supply of fossil fuels, heated debates around nuclear energy or heightened efforts to mitigate and adapt to climate change. Moreover, energy and the environment are increasingly a top priority of the European agenda and both issues lend themselves well to long-term planning and real transformative change.

In conclusion, Domenico Rossetti's book offers an innovative strategic view on economic, energy, research and environment policy in the EU. Beyond the traditional institutional tripod (Commission, Parliament and Council) and the formal consultation with civil society (industry, trade-unions and NGOs), this book illustrates the key role of scientists in supporting EU policy-making. Long-term policies are able to change the nature of the way we live today and act as a beacon for others around the globe.

Pascal Lamy

Director-General of the World Trade Organization

Former Member of the European Commission

ACKNOWLEDGMENTS

I wish to thank several people that helped me from the beginning of this fascinating work almost ten years ago and, first of all, Lara, my exceptional wife, for her affection and support during all this time. Drafting a book require time, a lot of time that I could have otherwise spent with her.

Secondly, many thanks to Prof. Jean-Marie Chevalier. This book is the result of a PhD thesis ("Doctorat d'État") in economics defended at the Paris Dauphine University *cum laude*. Prof. Chevalier, Supervisor of my thesis, always supported me in an enthusiastic way and with substantial academic advices and recommendations. I am grateful to Prof. Jan Horst Keppler for having chaired the PhD academic panel and to Dr. Dominique Finon and Prof. Paul Zagamé for their constructive comments, especially on the energy and economic modelling chapters.

Many thanks to Pierre Valette, a key actor in the evolution of the economic, environment and energy models as well as a "visionnaire" about the research challenges in Europe. I am also particularly grateful to Prof. Nigel Lucas for his careful reading of the text and for his precious comments.

I would like to thank to Emilie Menz, the Peter Lang Editor for her confidence; Prof. Vincent Dujardin and Prof. Amine Aait-Chaalal for the support of the Institut d'études européennes of the Catholic University of Louvain and to Dominique Charbit from the Paris "Centre de Géopolitique de l'Énergie et des Matières Premières" (CGEMP) for her practical day-to-day help; and Alice de Patoul for her excellent editing work.

The classes, talks and discussions with my professors and colleagues in the academic, research and institutional community during these last fifteen years have largely contributed to my knowledge of energy and innovation economics from one side and to legal and political aspects from the other side. Overall, my better understanding of European affairs is largely due the interactions with tens of people. In particular, I show regard for Jean-Michel Baer, Ezio Andreta, Bertrand Château, Pippo Ranci, Christian Franck, Pablo Fernandez Ruiz, Andrea Ricci, Patrick Criqui, Michel Dumoulin, Raffaele Liberali, Wolfgang Eichhammer, Sergio Romano, Catherine Mitchell, Etienne Cerexhe, Luc Soete, Jean-François Marchipont, Eberhard Jochem, Giuseppe Sfligiotti, Pantelis Capros, Michel Poireau, Christoph Böhringer, Fabrizio Caccia

Dominioni, Jacques Berleur, Giancarlo Tosato, Ignacio Pérez Arriaga, Alfred Voss, Alfonso Gonzalez Finat, Arturo Lorenzoni, Edward Steinmueller, Jacques Percebois, Nikos Kouvaritakis, Wiktor Raldow and Claude Thonet.

Thank you to my father for having nourished my European spirit from a young age, to my mother for having given me the inclination to write and to my sister Bénédicte for her attention.

This book is dedicated to my young and wonderful children Alessandro, Maria and Pablo, Europeans of tomorrow.

CHAPTER I

Setting the Scene

Introduction

In addition to the purely scientific and technological energy-related issues (such as the influence of energy consumption on global warming phenomenon or the development of energy technology), policy-makers must address the economic and social consequences of their choices. They can support their decisions with projections of socio-economic factors and evaluations of impacts as determined by quantitative tools, mainly economic and energy models.

Speaking about the applications of macro-econometric modelling in France, Paul Zagamé already underlined twenty years ago the important role of models in providing economic information: "Prévisions, variantes analytiques, scénarios complexes et études macro-sectorielles: de cette masse importante de travaux peut se dégager (...) le rôle important que les modèles ont joué dans le dispositif d'information économique"[1].

Recently and more directly concerning energy-related policies, Jean-Charles Hourcade, Mark Jaccard, Chris Bataille and Frédéric Ghersi highlighted that:

> Policy-makers are interested in a better understanding of the effectiveness and cost of policies (...). What technologies would serve this purpose, and how could or would the economy adapt in response to policy to achieve this end? Two contrasting modelling types have developed to answer these questions: conventional Bottom-up models and conventional Top-down models[2].

The focus of this book on economics and energy is justified for two reasons. Firstly because there is a relation between energy consumption and economic growth, as underlined by Jean-Marie Chevalier, "Il existe un lien fort mais complexe entre la consommation d'énergie et le déve-

---

[1] P. ZAGAME, *L'expérience française de modélisation macro-économétrique: bilan et perspectives*, Revue d'économie politique, No. 5, Paris, 1987.

[2] J.C. HOURCADE, M. JACCARD, C. BATAILLE, F. GHERSI, *Hybrid modelling: New answers to old challenges*, The Energy Journal, Special issue, IAEE, Cleveland, 2006.

loppement économique"[3]. Secondly, because most of the models that will be analysed in this book are either Top-down, i.e. economic-oriented, either Bottom-up, i.e. energy technology-oriented.

At the national-level (cf. *Commissariat Général du Plan* – now *Centre d'Analyse Stratégique* – in France or the Department of Energy in the US), European-level (European Commission) and international-level (International Energy Agency), modelling activities provide systematic and quantified information to decision-makers on the expected future according to various scenarios. Scenarios include a "reference case", together with variants appropriate to the issue under study, e.g. impacts of a climate policy on medium to long-term CO_2 emissions, consequences of a gradual nuclear phasing out on the energy mix, effects from an intensive renewable energy scenario on employment, impacts from a depletion of fossil fuel on the energy system, etc.

Without entering in the modelling formalisation (conception, variables, equations, time series, language), the objective of this book mainly consists in analysing the influence of results coming from scientists – in particular economists (Top-down modellers) and engineers (Bottom-up modellers) – on the European decision-making process.

This book does not enter in the debate where economic analysis leads to stronger or weaker regulation[4]. Confronted by various risks and challenges – especially when the future (10 to 50 years ahead) and the potential impacts (costs and benefits) are concerned – are (economic) researchers playing a role in European Union (EU) policy-making? This book aims to demonstrate the answer to this question by providing facts, figures and precise references based on five case studies of EU energy-related policies covering ten years (1995-2005).

Economic and energy modelling must cope with uncertainties that exist in any exercise that "looks ahead" (future) or that evaluates the potential impacts of a specific policy or measure. As Dominique Finon notes

---

3 J.M. CHEVALIER, *Les grandes batailles de l'énergie*, Gallimard, Paris, 2004. This complexity is well reflected in the conclusions of a study led by Jan Horst Keppler. Studying the relation between energy consumption and economic growth in ten developing countries, he concluded that China and India show a solid relationship while the other eight did not show a link between various indicators of *per capita* energy consumption and *per capita* income at the five per cent level of significance. J.H. KEPPLER, *Causality and cointegration between energy consumption and economic growth in developing countries*, in J.H. KEPPLER, R. BOURBONNAIS, J. GIROD (eds.), *The econometrics of energy systems*, Palgrave Macmillan, New York, 2007.

4 R. MORGENSTERN, *Economic analysis at EPA: Assessing regulatory impacts, Resources for the Future*, Washington, DC, 1997.

La combinaison d'hypothèses dynamiques sur les technologies, les comportements, la perception des raretés et la croissance conduit à un multi équilibre, c'est-à-dire à un éventail très vaste de futurs. Mais il ne reflète pas la réalité de l'incertitude globale sur la trajectoire d'ensemble, car les paramètres des systèmes technologique, économique et social interagissent entre eux pour opérer la sélection des futurs[5].

Even if the efficiency of the mathematical tools has been demonstrated in some sciences (like in physics) and the efficiency of informatics has been largely improved these last thirty years, this efficiency has not the same characteristics in economy. For example, in the energy field, it is still very difficult to anticipate the price of fuels, the energy demand or the behaviours of the agents in liberalised energy markets.

J.M. Chevalier highlights the weaknesses of the model results in the energy field.

Malgré la sophistication des outils utilisés, la prévision énergétique pure demeure un exercice impossible. On s'est tellement trompé dans le passé que l'on a aujourd'hui la certitude de ne pas savoir prévoir. On s'est trompé sur l'évolution des prix, le volume des ressources, le niveau attendu de la demande, les coûts estimés de telle ou telle forme d'énergie[6].

That is also why several authors prefer to represent uncertain factors through scenarios. As Peter Schwartz wrote:

Tout modèle économétrique repose sur une série d'hypothèses traitant de la manière dont fonctionne le monde. Le processus d'élaboration d'un scénario est particulièrement utile lorsqu'il s'agit de réfléchir aux conséquences des changements fondamentaux et des discontinuités qui mettent en cause ces hypothèses[7].

Nevertheless, this difficult nature of the economic and energy modelling does not reduce the important value – according to our initial postulate – of these studies for supporting EU energy-related policy-making. Their value lies in the systematic approach that can provide quantitative assessments of the impacts of policy measures and the factors that can affect or modify those impacts. Moreover, as Albert Strub underlined at the end of the 1970s: "The critical use of energy modelling can help, better than any intuitive effort, to narrow down the

---

5 D. FINON, *Prospective énergétique et modélisation – Identification de pistes de progression méthodologique*, Note au Conseil Scientifique de l'Institut français de l'Énergie, Paris, juin 2003. A shorter version of this paper can be found in D. FINON, *Prospective énergétique et modélisation de long terme: les voies de progression méthodologique*, Revue de l'Énergie, No. 553, janvier 2004.

6 J.M. CHEVALIER, *Les grandes batailles de l'énergie*, *op. cit.*

7 P. SCHWARTZ, *La planification stratégique par scénarios*, Futuribles, Paris, mai 1993.

multitude of possible future choices to those most promising economically and in particular to those most attractive to our society"[8].

At the EU-level, in the field of energy, some examples of recent issues justify the interest of this book: What is the impact of a EU-wide energy taxation? What medium to long-term target (for renewable energy sources, for energy efficiency, or for combined heat and power) has to be fixed? What is the most cost-effective or cost-benefit climate change policy? What monetary value has to be attributed to energy socio-environmental damages (external costs)?

Even within the chosen areas (taxation, energy and environment targets, etc.), the analysis reported here is selective: it covers mainly the quantitative tools, the EU policies in the energy-related field and the specific period since the middle of the 1990s to 2005. Many economic studies are broad in their scope and contribute to several of the aspects with which this work is concerned. Their outputs may be omitted in certain contexts simply in order to focus on the scope of the book, i.e. economic and energy modelling with EU decision-making process.

This book has the objective to detect the influence, if any, of the European energy economic research on EU policy-making. More specifically, the aim is to make an analysis of EU-decisions in energy-related fields and to underline the reasons of such choices, i.e. putting in evidence – and that is the main assumption – the role of economic research results in final EU policy decisions[9].

As Alan Manne, Richard Richels and John Weyant pointed out thirty years ago

> Energy policy is an interdisciplinary field. It involves economics, law, politics, engineering, resource geology, biomedical impacts, and environmental risk assessment – along with the methodologies that are already familiar to the operations researcher: optimization algorithms, simulations, decision analysis and econometric estimation. Unlike business applications in the private sector, it is exceedingly difficult to determine the impacts of these studies[10].

By answering to the question of modelling influence on the European policy-making in the energy-elated field, the following issues will be covered:

---

8 A. STRUB, *Energy systems analysis*, International Conference of Dublin, 9-11/10/1979, D. Reidel-Kluwer, ECSC-EEC-EAEC, Brussels and Luxemburg, 1980.

9 D. ROSSETTI di VALDALBERO, *Energy market regulation: Impacts of EU research*, in Ch. BÖHRINGER, A. LANGE, *Applied research in environmental economics*, ZEW-Physica-Verlag, Heidelberg, 2005.

10 A. MANNE, R. RICHELS, J. WEYANT, *Energy policy modelling: a survey*, Operations Research, Vol. 27, No. 1, 1979.

- EU energy research dealing with nuclear (fission and fusion) and non nuclear research, essentially fossil fuels, energy efficiency, renewables, fuels cells and hydrogen highlighting the economic research in energy also called "energy system" research or "energy strategic" research undertaken since the 1970s in Europe[11].
- Qualitative *versus* quantitative instruments, i.e. the scientific controversies on energy foresight taking into consideration the various tools for looking at the future(s) and for evaluating policies such as Scenario method, Delphi surveys, economic and energy models, or longer-term exercises like the ones developed by Shell or by the Intergovernmental Panel on Climate Change (IPCC)[12].
- Top-down and Bottom-up models with their economic or technological explicitness, with some critical issues (database, paradigms, assumptions, exogenous parameters, driving forces, invariants, "reference", policy scenarios, sociological aspects and technical and geographical representations), with long-term challenges and "backcasting" approach, and with the treatment of technological progress and climate change policies[13].

---

11 D. ROSSETTI di VALDALBERO, B. SCHMITZ, W. RALDOW, M. POIREAU, *European Union energy research*, Revue de l'Énergie, No. 576, Paris, March-April 2007.

12 G. BERGER, *L'attitude prospective*, Revue prospective, No. 1, 1958; J. DURAND, *Prospective, discontinuité et instabilité*, Futuribles, Paris, Automne 1975; D.H. MEADOWS, D. MEADOWS, J. RANDERS, W. BEHRENS, *Limits to Growth*, Universe Books, New York, 1972; P. WACK, *La planification par scénarios*, Futuribles No. 99, Paris, mai 1986; J. LESOURNE, Ch. STOFFAES, *La prospective stratégique d'entreprise – Concepts et études de cas*, Intersections, Paris, 1996; M. GODET, *The art of scenarios and strategic planning: Tools and pitfalls*, Technological Forecasting and Social Change, New York, 2000; K. VAN DER HEIJDEN, P. SCHUTTE, *Key questions for designing scenario applications*, Scenario and Strategy Planning, Vol. 1, No. 6, 2002; Shell, *Energy needs, choice and possibilities: scenarios to 2050*, London, 2001; IPCC, *Climate change, fourth assessment report*, Summary for policymakers, WMO/UNEP/IPCC, 2007.

13 C. BOEHRINGER, *The synthesis of Bottom-up and Top-down in energy policy modelling*, Energy Economics, No. 20, 1998; P. CAPROS, L. MANTZOS, P. CRIQUI, N. KOUVARITAKIS, A. SORIA RAMIREZ, L. SCHRATTENHOLZER, E.L. VOUYOUKAS, *Climate technology strategies*, Physica-Verlag, Germany, 1999; N. NAKICENOVIC, W.D. NORDHAUS, R. RICHELS, F.L. TOTH, *Integrative assessment of mitigation, impacts, and adaptation to climate change*, IIASA, Laxenburg-Austria, 1994; M. GRUBB, J. EDMONDS, *The cost of limiting fossil fuel CO_2 emissions – a survey and analysis*, Annual Review of Energy and Environment, Vol. 18, 1997; P. CRIQUI, L. VIGUIER, *The costs of CO_2 reduction and the impact of flexibility mechanisms in meeting the Kyoto targets*, Int. J. Global Energy Issues, 2000; D. FINON, *Prospective énergétique et modélisation de long terme: les voies de progression méthodologique*, *op. cit.*; J. GIROD, *Dynamic demand analysis and the process of adjustment*, in J.H. KEPPLER, R. BOURBONNAIS, J. GIROD (eds.),

- Reliability of model results analysing past projections (1985) with current statistical data and putting in evidence where and why some forecasts were extremely close or far to what really happened. This evaluation is based on the European "Energy 2000 study" that used the EFOM optimisation model and the MEDEE energy demand foresight model[14].
- Comparison of EU, international and US Bottom-up modelling-based studies of energy (WETO, WEO and IEO) allowing to "benchmark" the studies and to show the extent of agreement or difference among assumptions and results of the "Reference case" of the three studies covering the world with the 2030 timeframe[15].
- Science and society, research and policy interactions by underlying the EU *sui generis* decision-making process, the initiative power of the Commission, the specific EU Governance including the "comitology", the "scientific support to policies" highlighted in the European Research Area Communications, the use of expertise and the "Impact assessment"[16].

The econometrics of energy systems, *op. cit.*; L.R. KLEIN, *The Making of National Economic Forecasts*, Edward Elger, London, 2009.

14 J.F. GUILMOT, D. McGLUE, P. VALETTE, C. WAETERLOOS, *Energy 2000*, European Commission, Luxembourg, 1986; LAPILLONNE B., CHATEAU B., *The MEDEE model for long term energy demand forecasting*, Socio-Economic Planning Sciences, Vol. 15, issue 2, 1981; D. FINON, *Scope and limitations of formalized optimization of a national energy system*, in A. STRUB (ed.), *Energy models for the European Community*, IPC Science and Technology Press Limited, 1979; P.A. PILAVACHI, Th. DALAMAGA, D. ROSSETTI di VALDALBERO, J.-F. GUILMOT, *Ex-post evaluation of European energy models*, *Energy Policy*, Vol. 36, 2008.

15 European Commission, *World Energy Technology Outlook-2050 (WETO-H2)*, Luxembourg, 2006; International Energy Agency, *World Energy Outlook (WEO)*, Paris, 2006; US Department of Energy (EIA), *International Energy Outlook (IEO)*, Washington, 2006.

16 H. WALLACE, W. WALLACE, M.A. POLLACK, *Policy-making in the European Union*, Oxford University Press, 2005; F. CHALTIEL, *Le processus de décision dans l'Union européenne*, La Documentation française, Paris, 2006; J. RICHARDSON, *European Union – power and policy-making*, Routledge, 2006; Y. MENY, P. MULLER, J.-L. QUERMONNE, *Politiques publiques en Europe*, L'Harmattan, Paris, 1995; B. STEUNENBERG; C. KOBOLDT, D. SCHMIDTCHEN, *Comitology and the balance of power in the European Union*, International review of Law and Economics, 1996; W. WESSELS, *Comitology: fusion in action – Politico-administrative trends in the European Union system*, Journal of European Public Policy, Vol. 5, No. 2, 1998; European Commission, *European Governance – a White Paper*, COM(2001)428; European Commission, *Towards a European Research Area*, COM(2000)6; European Commission, *Making a reality of the ERA: guidelines for EU research activities (2002-2006)*, COM(2000)612; European Commission, *Communication on the collection and use of expertise by the Commission: Principles*

This book is multidisciplinary, treating different topics (energy, research, environment), but is most related to economics (models) and political science (decision-making). Although it also addresses EU legal and institutional issues – like the "new EU legislative culture" – and some engineering (Bottom-up) modelling results, this book focuses essentially on economic research and its role on the EU policy-making in the energy-related fields.

Up to now, this question has only been marginally explored on the theoretical side. This publication – also getting insights from experience – is therefore necessary to conceptualise the issues of scientific support to EU policies and eventually to improve the decision-making in Europe.

Methodology

The methodology starts from an experience-based established fact (*constat*); in the European Union, a policy initiative succeeds in its implementation if it receives a three-level support:

- If it responds to immediate economic and social needs, i.e. if the initiative is economically feasible and socially acceptable at a certain moment under specific conditions. It has often been mentioned in the literature that the conjunctural factor (*facteur conjoncturel*) plays a key role in the European construction. The role of "stakeholders" – of the representation of economic and social interests – is particularly relevant here[17].
- If there is adequate consensus among the major EU political actors, i.e. if the initiative can be subject to a broad agreement from Member States (Council), the European Parliament and the European Commission. The role of Ministers, Deputies and Commissioners as well as political parties and national and EU administrations are determinants[18].

---

and guidelines – Improving the knowledge base for better policies, COM(2002)713; European Commission, *Communication on Impact Assessment*, COM(2002)276.

17 M. ARTIS, F. NIXSON, *The economics of the European Union*, Oxford University Press, 2007; J.-C. DEFRAIGNE, *Economie européenne et politique*, L'Harmattan, Paris, 2004; A. LAMFALUSSY, L. BERNARD, A. CABRAL, *The Euro-zone: A new economic entity?*, Bruylant, Bruxelles, 1999; S. DESSELAS, *Un lobbying professionnel à visage découvert*, éd. du Palio, 2007; J.-D. GIULIANI, *Marchands d'influence*, Seuil, Paris, 1991; H. MICHEL, *Lobbyistes et lobbying de l'Union européenne*, Presses universitaires de Strasbourg, 2006; J. GREENWOOD, *Interest representation in the European Union*, Palgrave, 2007.

18 H. WALLACE *et al.*, Policy-making in the European Union, *op. cit.*; F. CHALTIEL, *Le processus de décision dans l'Union européenne*, *op. cit.*; J. RICHARDSON, Eu-

- If – in the energy and environment fields – the scientific community is in accord, i.e. if the initiative gets a large and shared scientific consensus. The role of scientists, mostly academics – but also from research centers, consultant companies, think tanks or sometimes from private firms – is crucial in the up-stream EU decision-making process.

The two first "supports" are largely studied from an economic, legal, political science, sociological or historical perspective. They are directly or indirectly part of the main master or doctoral courses delivered by the "EU institutes" existing nowadays in most European universities or by specialised academic institutions like the College of Europe in Bruges, the European Institutes from the Catholic University of Louvain or the Brussels University or the European University Institute in Florence. Several European research networks, Brussels-based Think-tanks, or Foundations also largely deal with these two supports like the Centre for Economic Policy Research (CEPR), the Centre for European Policy Studies (CEPS), the European Policy Centre (EPC), the European Centre for International Political Economy (ECIPE), the Brussels European and Global Economic Laboratory (BRUEGEL), the Friends of Europe, the Madariaga or the Robert Schuman Foundations.

De facto, these two "internal" supports (economic and social from one side, and political from the other side) have largely been studied by several academics in the whole range of social sciences[19].

The last policy support, the role of scientists in EU policy-making has essentially been tackled from a "Risk governance" perspective and mainly in relation to the influence of experts in natural sciences or medicine in human health, agro-food sectors or environment. Several publications concerned for example the role of scientists in EU policy issues like genetically modified organisms or mad cow disease (BSE crisis)[20].

---

ropean Union – power and policy-making, *op. cit.*; L. CARTOU, J.-L. CLERGEVIE, A. GRUBER, P. RAMBAUD, *L'Union européenne*, Dalloz, Paris, 2006.

19 The term "internal EU support" is used as another important support to EU initiatives comes from "external" forces, e.g. USA and the Soviet Union during the Cold War. Several academic papers and seminars focused on the international relations and geopolitical aspects influencing the EU integration like those from the "Institut Français des Relations Internationales" (IFRI), the UK Royal Institute of International Affairs – Chatham House – or the "Italian Istituto Affari Internazionali" with their respective publications (*RAMSES*, *International Affairs Journal*, *IAI Quaderni*). P. BONIFACE, *Les relations internationales de 1945 à nos jours*, Dalloz, Paris, 2005.

20 H. TORGERSEN, *The real and perceived risks of genetically modified organisms*, EMBO, 2004; C. JAEGER, O. RENN, E. ROSA, T. WEBLER, *Risk, uncertainty and rational action*, Earthscan, London, 2001; N. NEUREITER, *Challenges of providing scientific advices to governments*, JRC, Ispra, 26/10/2008.

Regarding genetically modified (GM) products, the Commissioner in charge of Agriculture said in 2008:

> I am certainly not arguing that we should authorise imports of GM products which science has told us to reject. But where science has given a product a clean bill of health, that fact must be paramount as we follow the authorisation procedure[21].

Food safety has been a subject particularly addressed these last few years in the EU. There is evidence that "Scientific advisory committees in Brussels tend to reflect approaches of national governments", as pointed out by Christoph Srünck. To manage this distortion, a European Food Safety Authority has been established[22]. Its primary responsibility is to provide independent scientific advice on all matters with a direct or indirect impact on food safety.

Environment is a subject where the need for and legitimacy of scientific support has been recognised since the Single European Act in 1986 and the establishment of one of the first European specialised Agency, the European Environment Agency in 1990. The Commission Climate Change "Adaptation" Green Paper clearly stated that "Sound scientific results are paramount in the development of climate policy"[23].

Concerning social scientists influence on EU decision-making, the academic literature is mostly limited to their role as "experts" in the so-called "Comitology", in different "working expert groups" and in the evaluation of European programmes and projects[24]. Frédéric Allemand

---

21 M. FISCHER BOEL, *Food, feed or fuel: a measured policy on agricultural markets*, Meeting with Bundesvereinigung der Deutschen Ernährugsindustrie, Berlin, 18 January 2008.

22 Ch. SRUENCK, *Why is there no mad cow disease in the United States? Comparing the politics of food safety in Europe and the U.S.*, Working Paper, Institute of European Studies, University of California, Berkeley, December 2001; H. VOS, F. WENDLER, *Food safety regulation in Europe*, Intersentia, Antwerpen-Oxford, 2006; J. BOURINET, F. SNYDER, *La sécurité alimentaire dans l'Union européenne*, Bruylant, Bruxelles, 2003.

23 European Commission, *Green Paper, Adapting to climate change in Europe – options for EU action*, COM(2007)354.

24 B. STEUNENBERG; C. KOBOLDT, D. SCHMIDTCHEN, *Comitology and the balance of power in the European Union*, International review of Law and Economics, 1996; W. WESSELS, *Comitology: fusion in action – Politico-administrative trends in the European Union system*, *op. cit.*; T. CHRISTIANSEN, T. LARSSON, *The role of committees in the European-Union policy-process*, Edward Elgar, 2007; U. PUETTER, *Providing venues for contestation: The role of expert committees and informal dialogue among ministers in European economic policy coordination*, Comparative European Politics, Vol. 5, Palgrave Macmillan, 2007.
The role of scientists is also large in the evaluation of EU programmes. For example, in the EU Framework Programme (€ 54 billion for the seventh RTD FP covering the 2007-2013 period), the funding decisions are based on peer review of research pro-

highlights that with the increasing technicity of the subjects, the experts committees become stronger and stronger and the role of ministers is more and more limited:

> La technicité des sujets reporte les discussions au sein des comités d'experts, réduisant les formations ministérielles à une simple fonction de chambre d'enregistrement. Les Ministres lisent les propositions préparées en amont[25].

The interaction among social sciences, especially economics, energy and policy-making in Europe has never been analysed in an in-depth way. The methodology developed here focuses on a specific field – energy –, on a specific policy system – European Union, and on a specific period – 1995-2005.

Figure I.1: Economic and social support, political support, and scientific support needed for the adoption of EU policies

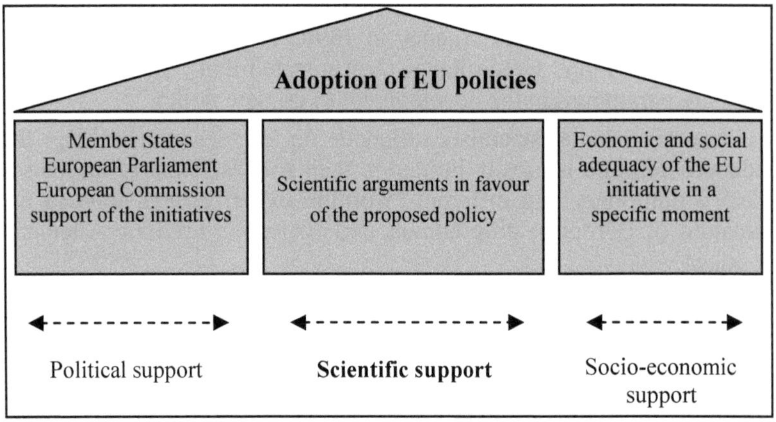

Based on a theoretical background and on five case studies, the methodology specifically tackles the issue of economic and energy modelling with EU decision-making from different approaches:

- Description of EU energy research by statistics and history (Chapter II);

posals. External experts (mainly scientists called "evaluators") are at the core of the system. Over 50,000 evaluators – including thousands of social scientists – were registered in the database by the end of the sixth Framework Programme (2006). About 4,000 independent experts are engaged every year by the European Commission to evaluate about 16,000 proposals per year.

25 F. ALLEMAND, *L'Union pour la Méditerranée: Pourquoi? Comment?*, Fondation pour l'innovation politique, Paris, Juin 2008.

- Analysis of scientific controversies on quantitative and qualitative tools for energy foresight by literature review and examples of international studies (Chapter III);
- Assessment of potential methodological improvements on Top-down and Bottom-up models by studies and experience on EU energy economic research activities (Chapter IV);
- Description of EU modelling tools by illustrative representation of the economic and energy systems (Chapter V);
- Demonstration of the role of economists (modellers) in EU energy-related policies through five case studies (Chapter VI);
- Validation of modelling results by *ex-post* evaluation and by comparison of EU, international and USA energy outlook reports (Chapter VII);
- Identification of the role of experts in EU policy-making by legal analysis, *de facto* implementation means and deduction from previous chapters (Chapter VIII).

The last part of the work contains a final discussion and concluding messages (Chapter IX) as well as a meta-conclusion on the future of modelling in the EU decision-making process.

In drafting this book, the author considers that his professional life at the European Commission and his consequent experience of EU decision-making process have been an asset. It has facilitated access to relevant documents (cf. Commission staff working documents or preparatory studies made for the "Impact Assessment" of a specific initiative), but also has engendered a personal experience of the influence of scientists in EU energy-related policies.

The author considers that his situation allowed him to help to theorise some important aspects – but often neglected – of EU decision-making, i.e. the role of scientists and in particular of economic and energy modellers in up-stream EU policy-making. In addition, this publication is a further step for increasing the transparency of the EU decision-making process, in particular in its crucial initial political phase.

Although involved in the development of EU energy economic research and in EU-energy-related policies, the author has taken care of being impartial in the assessment of the information analysed in this book. In addition, in order to respect his "devoir de réserve" as European civil servant, the author has explicitly addressed EU policy cases up to 2005 (and not later) while this book is published in 2010.

Structure of the work

The first step aims to highlight the EU "Economic research in energy" also called "Strategic energy research". What topics and what budgets were dedicated to scientists, mainly energy economists, from the first to the sixth EU RTD Framework Programme. In this chapter, a complete state of the art of the purely EU energy RTD (i.e. not the Member States but only the Community Framework Programmes priorities and expenditures) is made. A distribution among the energy RTD topics and linked budgets (nuclear fission, nuclear fusion, and non nuclear energy) is provided. This work has never been done and up to now, no statistics existed per energy sub-topics at the EU-level. During these last two decades, it should be remembered that the budget of the Framework Programmes represented about 4% of the European Community's budget and approximately 6% of the EU's public civilian research investment.

The RTD expenditures dedicated to economic research in energy are highlighted in the six Community RTD Framework Programmes (1984-1987, 1987-1991, 1990-1994, 1994-1998, 1998-2002 and 2002-2006). This part of the work requires an in-depth analysis of the EU funding in the energy field. As the RTD Framework Programmes have only budgetary divisions in terms of "macro topic" such as energy, biotechnology, information and communication technologies, etc. a "project by project" evaluation is performed. This is the only way to get a real picture of the EU priorities and funding sub-division in the energy field. The breakdown is made between nuclear (fusion and fission) and non nuclear RTD and among non nuclear fields, i.e. essentially economic research on energy, fossil fuels, energy efficiency, CO_2 capture and storage, fuel cells, hydrogen and storage, and renewables – with a distinction among wind, photovoltaic, biomass, geothermal, solar, ocean, hybrid renewables and distributed generation.

The economic research on energy launched in the 1970s assumes today the role of "scientific support to policies". Since 2000, this has been formalised in two Commission Communications concerning the *European Research Area*: "Developing the research needed for political decisions..." and "Support for policy-making and European scientific reference system..."[26]. In addition, these research activities supporting EU policies were a precursor of the formal *Impact Assessment* "which identifies the likely positive and negative impacts of proposed policy

[26] European Commission, *Towards a European Research Area*, *op. cit.*; European Commission, *Making a reality of the ERA: guidelines for EU research activities (2002-2006)*, *op. cit.*

actions, enabling informed political judgements"[27]. Since 2003, an Impact Assessment is obligatory for all Commission major initiatives; the intention is to improve the quality and coherence of the policy development process ("better regulation" as indicated in the 2001 Commission White Paper on European Governance)[28].

In this first part of the book, EU energy research priorities and budgets are detailed. The share of socio-economic research in energy (or, more recently called research dedicated to policy support) is about 3-4% on non nuclear energy research. The main topics addressed concern the:

- Development of tools for energy strategy (modelling, forecasting and new paradigms for a lower carbon energy sector);
- Methods for the economic and environmental assessment of energy production and consumption including the monetary valuation of externalities;
- Elaboration of scenarios for energy supply and demand technologies and their interaction, and the analysis of cost effectiveness (based on full life cycle costs) and efficiency of all energy sources;
- Social acceptability, behavioural changes and international dimension related to sustainable energy RTD.

The majority of the EU energy economic research projects makes the links between energy and environment and addresses the issues of natural resources, economic growth and social needs, including both market competition and environmental constraints, Top-down and Bottom-up approaches. Among the main topics covered by the EU energy economic research, it is worthwhile mentioning the Economics, Energy and Environment (E3) models and the energy external cost-related activities. The budget allocated to EU energy economic research is estimated at around € 3-4 Million per year.

This first part of the book is original, but mainly descriptive; it provides the statistics and the historical evolution of EU energy RTD essentially based on European Commission sources.

Chapters III and IV ("Scientific controversies on energy foresight" and "Economic and energy modelling") are analytical and theoretical. Based on the academic literature on foresight and economic/energy modelling, these chapters:

- Aim to highlight the scientific debate on quantitative and qualitative tools dealing with energy futures and with the evaluation of impacts of potential policies and measures (scenarios);

27 European Commission, *Communication on Impact Assessment*, *op. cit.*

28 European Commission, Communication, European Governance – a White Paper, *op. cit.*

- Synthesize the main strengths and weaknesses of these foresight and modelling tools also largely taking on board EU research in this field;
- Suggest some potential methodological improvements like a better dialogue between economists and technologists, between micro-economic realism and macro-economic completeness and between energy supply and demand sides

Starting from illustrative energy futures studies, such as those performed by the European Commission, the International Energy Agency, or the US Department of Energy, these chapters provide a new conceptual framework and theoretical classification in terms of:

- Quantitative (model-oriented) or qualitative (Delphi-oriented) approaches;
- Stakeholder (participatory) or expert-based works;
- Socio-political wide or techno-economic problem-solving scopes.

Most of the methods and tools dealing with the energy futures are addressed and discussed and the main critical points are highlighted. Two specific concerns are particularly underlined: making forecasts of the long-term (about 100 years) and technological progress (learning curves) especially in relation to policies for the mitigation/adaptation to climate change.

Several results from EU research projects quoted in Chapter II are given in these Chapters III and IV.

Chapter V is dedicated to the description of the main quantitative tools, essentially Economics, Energy and Environment (E3) models and methodologies developed by European researchers in the framework of the EU economic research in energy (cf. Chapters II, III and IV). These tools and methodologies are covering the world and/or the EU, are oriented towards the macro-economic issues or detailing the energy sector – sometimes focusing on specific issues like renewable energy sources and rational use of energy – or are quantifying the social and environmental damages, i.e. the energy external costs. The basic schemes of the EU macro-economic model (GEM-E3), the World and EU energy models (POLES and PRIMES), the EU energy softwares for renewables (SAFIRE) and for rational use of energy (MURE) and the EU external costs accounting framework (EXTERNE) are presented in this chapter.

Chapter VI is the empirical part of the book. Based on five case studies in the energy-related field, an assessment of EU policies (Commission "Communications", Commission proposals for directives, European Parliament and Council directives) is accomplished. The analysis addresses in particular the "origin of the figures" such as the impact of EU energy taxation, the cost of Kyoto and the economics of the EU emission trading system, the target for renewable electricity, the potential for energy efficiency, or the value of the electricity external costs. A particular attention is dedicated to the analysis of the Official Journal of the European Communities, the Commission staff working documents and the EU studies. Concretely, the following five EU case studies (political initiatives and decisions) in the energy-related field under the period 1995-2005 are analysed:

- The Community framework for energy taxation and in particular the proposal for a directive restructuring the Community framework for the taxation of energy products;
- The European Union Climate Change policy and in particular the Green Paper and the proposal for a Directive on greenhouse gas emission trading within the EU;
- The promotion of from renewable energy sources in Europe and in particular the Directive on the promotion of electricity from renewable energy sources;
- The Energy efficiency in the European Community;
- The measurement and the internalisation of external costs in Europe and in particular the Community guidelines on State aid for environmental protection.

This chapter makes the link between theory and practice. It reveals how EU energy economic research (Chapter II) and in particular its quantitative tools (Chapter V) are directly used in the preparation of EU policy-making in the energy field. It demonstrates by facts, figures and references the weight of scientists, mainly economists, in the up-stream EU decision-making.

After the case study demonstration of the impact (*ex ante*) of economic research work (quantitative tools) on EU decision-making (use of model results in EU policies), Chapter VII deals with an *ex-post* evaluation of an energy futures exercise (cf. Chapters III and IV) led in the middle of 1980s for the EU at the end of the century. In fact, to date there have been few careful comparisons between the realized and the forecast effects in the energy system although it is useful to make some evaluation of the accuracy of prospective estimates. Comparing the results of the (1985) European Commission study "Energy 2000" – performed with the so-called MEDEE and EFOM models – with current Eurostat data aims to highlight the similarities and the differences between the projection results and the real data in terms of overall energy consumption and in terms of individual energy source and technology.

The model "benchmarking" is also completed through a comparison of three Bottom-up modelling-based studies published by the European

Commission (EC-WETO), the International Energy Agency (IEA-WEO) and the US Department of Energy (US-IEO). Comparison of the assumptions and results by 2030 demonstrate where high concordances or relatively large differences exist among world energy models (POLES, WEM and SAGE).

Looking at the "origin of the figures" provided in the EU policies (case studies), Chapter VIII concerns the interaction of science and society, more particularly of research results and policy-making at the EU-level.

The specificities of the decision-making process at the EU-level is underlined, especially the "techno-political" role of the Commission. The increasing role of the "soft law" including the Communication, Green and White papers is recognised. In these official, but not legally binding documents, the role of the scientists – in particular economic modellers for energy-related policies – is obvious and important although too often neglected in the literature.

The role of experts in up-stream decision-making is not only captured in several EU legal acts (from the European Treaty to the guidelines on the use of expertise) but also in practice through the so-called "Comitology" and Impact Assessment.

Several messages are underlined in the last concluding chapter. They are related to EU energy and research policies; to qualitative and quantitative tools for energy policy analysis; to economics, energy and environment modelling and its links (interactions) with EU decision-making; to the "power of scientists" in up-stream EU policy-making. Finally, the author highlights some future modelling challenges with respect to EU policy-making.

Although the author recognises the importance of "policy drivers" in the development of new modelling tools, e.g. the fact that the oil crises in the 1970s, the sustainable development concept in the 1980s and the liberalisation of the energy markets in the 1990s have given a great impetus in the generation of new models, this subject is not covered in this book.

CHAPTER II

Energy Research in the European Union

Summary

The objective of this chapter is to review in detail the development of energy research within the first six Community RTD Framework Programmes. The first one (FP1) covers the period 1984-1987; the second one (FP2) covers the period 1987-1991; the third one (FP3) covers the period 1990-1994; the fourth one (FP4) covers the period 1994-1998; the fifth one (FP5) covers the period 1999-2002 and the sixth one (FP6) covers the period 2002-2006[1].

European Research and Technological Development (RTD) activities have initially been carried out on a case by case approach, within each of the three Treaties which have established the foundations of the European Union: the European Coal and Steel Community (ECSC), the European Atomic Energy Community (EAEC also called Euratom) and the European Economic Community (EEC).

Since 1984, these RTD activities have been integrated into comprehensive and multiannual schemes: the Framework Programmes (FP) that cover all research fields considered as European priorities. This approach was meant to provide better visibility and transparency in the decision-making process, including for the choice of thematic priorities.

Energy RTD has been included into the successive FPs:

- since 1984 for its research part (covering nuclear and non nuclear energy);
- since 1990 for the demonstration component of the non nuclear area.

Despite the complexity of the legal and institutional landscape, the quantitative and qualitative analysis of the European Community (EC) energy RTD, provides a series of messages:

- while energy research has been the major RTD area before the creation of the FPs in the mid 1980s, its share has progressively de-

[1] Most of this Chapter has been published in D. ROSSETTI di VALDALBERO, B. SCHMITZ, W. RALDOW, M. POIREAU, *European Union energy research*, *op. cit.*

creased with respect to other technologies going from about 50% of Community RTD budget in FP1 to 10% in in FP6;

- within the growing RTD budget of the EC, the share of energy in relation to other topics like life sciences, information and communication technologies, new materials, has decreased rather quickly in relative terms from 1984 to 1990, and has then remained rather stable over the last 15 years (1990-2006);
- the focus of energy research has been primarily on issues dealing with public safety and security (first in nuclear with nuclear safety, safeguards etc.; then after, the first oil shock, extended to improve European security of supply through activities in the non nuclear field); secondly with long term and risky endeavours such as fusion; thirdly with support of new and emerging technologies such as renewable energy sources, fuel cells and hydrogen; lastly with the economic research in energy also called strategic energy research supporting EU policies;
- with about ECU/EURO (€) 1 Billion current per FP, the nuclear energy research budget (Euratom) has been kept more or less constant but its relative share fell from around 25% in FP1 to 6% in FP6;
- the non-nuclear energy RTD budget has been subject to substantial fluctuations, from € 830 Million in FP1 to € 120 M in FP2, from € 260 M in FP3 to more than € 1 Billion in FP4 and FP5 and to € 800 M in FP6; these variations are moderated if demonstration of energy RTD – only included in the RTD FP since FP4 – are considered in the period going from 1984 to 1994, i.e. corresponding to the time-schedule of the first three FP.

From 2000, the European Research Area (ERA) has become the basic policy reference for research in the EU aiming at reinforcing the knowledge economy and society and promoting innovation, competitiveness, employment and sustainable development. ERA introduces new instruments in European research and FP6 is its main legal, planning and financial tool.

Energy RTD has always been a substantial part of common EC research, both in the nuclear and non-nuclear energy field. Nevertheless, even if EC energy RTD remains considerable today, taking into account inflation and successive EC enlargements, the RTD effort in the field of energy has significantly decreased over time.

Economic research in energy should not be confused with the purely "socio-economic research" – like TSER at the end of the 1990s – which dealt these last ten years essentially with science and technology policy options, education and training, competitiveness, migration, social exclusion and social integration. Until now there were no reliable information and statistics able to provide an appropriate quantitative monitoring of the economic issues addressed in the different FP in the energy sector. In the energy field, a very large part of the economic research in energy is dedicated to the development of tools, essentially models that are used for evaluating policies and measures, and to the development of an accounting framework for the quantification of the energy external costs. These models and accounting framework will be described in Chapter V. The budget dedicated to economic research in energy at the EU-level is about € 4 Million per year since 1990.

European research and energy research in the six RTD Framework Programmes

The analysis carried out in this chapter deals with the Community's RTD actions and not with the specific RTD activities of the Member States. It will focus on the major EC instrument, i.e. the Framework Programmes from 1984 to 2006.

Energy research activities have been carried out under the aegis of each of the three Treaties, which are at the foundation of the European Union[2]:

- the European Coal and Steel Community (Treaty of Paris, 18th of April 1951) largely cover the coal research (Article 55 of the ECSC Treaty – expired in 2002 – was the basis of the EU RTD policy);
- the European Atomic Energy Community (EAEC or Euratom) for nuclear research (Treaty of Rome, 25th of March 1957) and especially Articles 4 to 11;
- the European Economic Community (EEC) also for non nuclear energy (NNE) research (Treaty of Rome, 25th of March 1957) and especially Articles 35 and 308.

On 14 January 1974, the Council presented four resolutions in the field of science and technology, officially inaugurating a non-sectoral R&D Community policy[3]. These resolutions were inspired by:

- the perception of a "technology gap" between Europe and USA; this will be particularly demonstrated before the launching of the first

2 N. LUCAS, *Energy and the European Communities*, Europa Press, 1977.

3 Council resolutions concerning (1) *the coordination of national policies and the definition of projects of interest to the Community in the field of science and technology*; (2) *the participation of the European Communities to the European Science Foundation*; (3) *a first action programme of the European Communities in the field of science and technology*; (4) *an action programme of the European Communities in the field of forecasting, evaluation and methodology*, OJEC, C 7, 29/1/1974.

Framework Programme and the ESPRIT programme related to information technologies[4].

- the "relance" of the European Summit in The Hague (1-2 December 1969) in which RTD appears as a new and important building block for the construction of Europe. The will to develop a common European policy in the field of science and technology will be also expressed by the Heads of State and Government at the Paris European Summit (19-20 October 1972).
- the unification of the Commissions and Councils of the three Communities (ECSC, EEC, EAEC), effective in 1967, was a precondition for a unified research policy.

Title XV EC provided a new and explicit basis for research and technological development policy, based on a five-year RTD Framework Programmes.

The European R&D budget figures mentioned in this chapter are expressed without distinction in current ECUs or EUROs (€). The reconstruction of the European R&D budget takes into account the various enlargements of the European Community since the first RTD FP in 1984: to Portugal and Spain in 1986, to Austria, Finland and Sweden in 1995 and to Cyprus, Czech Republic, Estonia, Hungary, Latvia, Lithuania, Malta, Poland, Slovakia and Slovenia in 2004. These figures correspond to the amounts decided within the legal decisions that have established successive Framework and Specific programmes over that period.

Parallel changes occurred also in the decision-making process when:

- European Parliament increased strongly its influence through the co-decision process on non nuclear energy research (after the 1992 Maastricht Treaty); concretely, from 1993 co-decision by Council and Parliament for the adoption of the EU Framework Programmes was introduced and the aim of EU RTD policy redefined. FP4 will be the first Framework Programme emerging from a European Parliament and Council decision. Article 7 of the Euratom Treaty was left unchanged.
- Unanimity of Member States was abandoned in non nuclear energy research after the 1997 Amsterdam Treaty (it is still required for nuclear in Euratom based activities). *De facto*, in 1999 the requirement for Council unanimity to adopt the Framework Programme was replaced by qualified majority; this will be firstly the case with FP6.

[4] European Commission, *Towards a European strategic programme for research and development in information technologies*, COM(82)287; and Council decision on a *European strategic programme for research and development in information technologies (ESPRIT)*, OJEC, L 67, 9/3/1984.

The Euratom Treaty was left unchanged, and the Euratom FP is adopted unanimously by the Council following a single reading by the European Parliament. Specific programmes within the FP are adopted by qualified majority in Council, following simple consultation of the EP

The aim of EU RTD policy is to strengthen the scientific and technological basis of European industry and to encourage it to become more competitive at international level. Promoting industrial competitiveness was central to the FP as defined by the Member States in the 1980s. This key provision was amended in the Maastricht Treaty by adding the phrase while promoting all the research activities deemed necessary by virtue of other chapters of this Treaty.

Article 164 specified: "In pursuing these objectives, the EU shall carry out the following activities, complementing the activities carried out in the Member States":

- implementation of research, technological development and demonstration programmes, by promoting cooperation with and between undertakings, research centres and universities;
- promotion of cooperation in EU research, technological development and demonstration with third countries and international organisations;
- dissemination and optimisation of the results of activities in EU research, technological development and demonstration;
- stimulation of the training and mobility of researchers in the EU.

Article 165 provides that "The EU and Member States shall coordinate their activities so as to ensure that national policies and EU policy are mutually consistent."

The main instruments of EU RTD policy are the multi-annual Framework Programmes (FPs), which set objectives, priorities and financial budgets for a period of several years.

The EU achieves its RTD objectives by two ways:

- direct research activities projects carried out by the EU Joint Research Centre (JRC) located at Ispra (Italy), Geel (Belgium), Petten (Netherlands), Karlsruhe (Germany) and Seville (Spain);
- indirect action (EU RTD projects) carried out by groups of researchers coming from industries, laboratories and universities essentially from the Member States and generally financed by the EU to a maximum of 50%.

At the international level, Europe lags USA and Japan in terms of R&D intensity, i.e. R&D effort in comparison to GDP[5]. If the current trend persists, China could catch up the EU by 2010. Compared with its main competitors on the world market, R&D spending is relatively low (1.9% of GDP for R&D in the EU, 2.6% in the US and 3.1% in Japan in 2005)[6]. The Heads of State and Government agreed at the Barcelona European Council (15-16 March 2002) to fix a target of 3% of GDP dedicated to RTD by 2010[7].

Currently, the EU research spending (from the FP6) amounts to about 6% of the total public civil research in the EU Member States[8]. Universities across the EU employ about 37% of researchers, compared to around 15% in the US and 26% in Japan[9]. The number of researchers in full time equivalent (FTE) per thousand labour force amounted to 5.4 in the EU in 2003, compared to 10 and 9 in Japan and the US. Of the estimated total of 1,180,000 researchers (FTE) in the EU-25 in 2003, about 50% were employed in the business sector. This compared to some 68% in Japan and about 80% in the US[10].

In addition to the Framework Programme, in order to promote R&D, the EU uses:

- political instruments (like the overall Barcelona objective of 3% of R&D in the EU by 2010),
- financial instruments and fiscal incentives (like the loans from the European Investment Bank or the structural funds),
- legal instruments (like EU patent, European Economic Interest Grouping – EEIG, worldwide cooperation agreements, Joint undertakings or intellectual property right conventions) and
- mobility instruments (like accreditation methodology for exchange of students and teachers ("Bologna" and the European Credit Transfer System – ECTS).

---

5 European Commission, *Third European report on Science and Technology indicators*, OPEC, Luxembourg, 2003.

6 European Commission, Staff working document accompanying the Green Paper, *The European Research Area: New perspectives*, SEC(2007)412.

7 European Commission, *More Research for Europe: Towards 3% GDP*, COM(2002)499.

8 European Commission, *Participating in European research*, Brussels, 2004.

9 European Commission, *Staff working document accompanying the Green Paper on the European Research Area: New perspectives*, *op. cit.*

10 European Commission, DG Research, *Key Figures 2005*, Brussels, 2006.

Figure II.1: RTD intensity in the EU, Japan, US and China (Gross domestic expenditure on RTD as percentage of GDP)

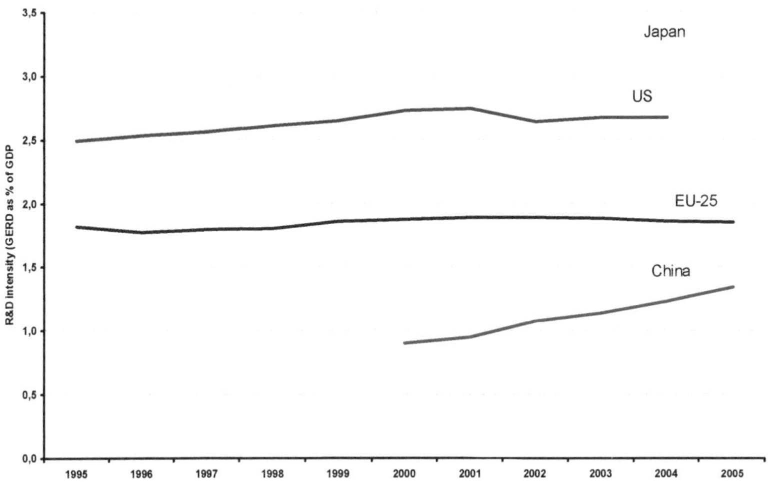

Among the EU Member States, there is a great difference of efforts in terms of RTD. Public and private spending of Sweden and Finland is at more than 3% of GDP, the Southern and Eastern European countries spending are at less than 1.5%, and UK, France and Germany spending are between 1.5% and 2.5%.

Two thirds of all European RTD activities are carried out by industry. According to official RTD statistics for the EU-27, expenditure on R&D is distributed as follows: business (BERD) represented € 128 Billion in 2005 (64% of total), universities and higher education research and development spending (HERD) € 44 Billion (22%) and governmental research organisations spending (GOVERD) € 26 Billion[11].

In addition to insufficient RTD funding in many Member States, the EU research framework conditions (e.g. fiscal incentives for research, protection of intellectual property, availability of venture capital, etc.) are sub-optimal and not conducive to sufficient research or to its exploitation via new products, processes and services of new knowledge. Research and innovation policies are also pursued largely independently (with varying legislative, regulatory and financial structures) leading to a governance failure characterised by poor integration and coordination between various levels (regional, national, EU) and by a sub-optimal allocation of resources.

11 OECD, *Main science and technology indicators*, Paris, 2006.

Today, in their national policies, Member States generally recognise policy developments at the European level as factors that have to be taken into account in national policy making, with reference being made most often to the general objectives of the Lisbon strategy and the 3% objective[12].

Figure II.2: RTD intensity in the EU Member States (Gross domestic expenditure on RTD as percentage of GDP)

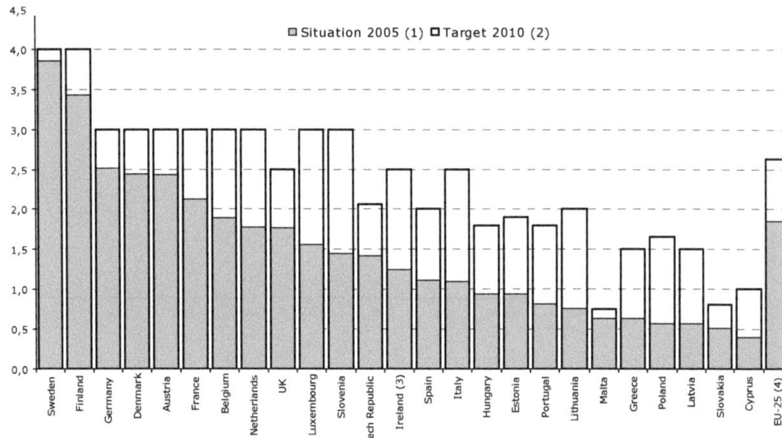

(1) Greece: 2003; Belgium, Italy, Malta, Netherlands, Slovenia, and UK: 2004. (2) Portugal: 2008; Ireland: 2013; UK; 2014. (3) Ireland: The target is 2.50% of gross national product in 2013. (4) EU-25: The EU-25 R&D intensity for 2005 was estimated by DG Research. EU-25: The EU-25 R&D intensity for 2010 results from the aggregation of the set targets set by the Member States (including estimated targets for Ireland, Portugal and the UK). Member States have been ranked according to the current level of R&D intensity from left to the right.

The picture of EU energy RTD provided in this chapter is consequently influenced by four factors:

- As *public RTD*, it reflects more the priorities linked to public interest and to long term or risky options than private RTD which does not intervene in areas such as nuclear fusion, or does it at a much lesser extent like in the field of renewable energy sources;
- As *European RTD*, it complements national activities, which are substantially more important in quantitative terms (as an average 2 to 3 times more in the energy field);
- As regards its implementation, the bulk of the budget is used to support trans-national activities, on a *shared cost* basis, where the

[12] European Commission, Green Paper, *The European Research Area: New perspectives*, *op. cit.*

EC bears between 30% – in demonstration projects – and 50% – in research projects – of the total cost. This means that altogether EC programmes influence directly or strongly a turnover of research and demonstration activities 2 or 3 times bigger than the sheer budget of the EC. The only exception to this shared-cost approach is represented by the Joint Research Centre which benefits from direct funding for the execution of research missions entrusted by the Treaties (especially in the nuclear field);

- As to the degree of *integration* achieved, a distinction must be made between fusion, fission and non nuclear energy. European fusion energy research has a strong programmatic rather than project-oriented character (fission and non nuclear energy) and is carried out with the Member States (and Switzerland) in a fully integrated and co-ordinated framework.

Evolution of European Union RTD budget

Before the RTD Framework Programmes were established in 1984, the three Communities had already started several activities of sectoral research: the ECSC Treaty provided for research on coal and steel; the EEC Treaty permitted research in agriculture; and nuclear R&D was a fundamental policy of Euratom. The 1974 Council Resolutions will allow the coordination of national policies and projects of interest to the EU in the fields of science and technology.

As shown in the Figure II.3, the global Community RTD budget has grown from about € 4 Billion in FP1 (1984-1987) to € 17.5 Billion in FP6 (2002-2006)[13]. The FP7 has a budget of over € 53 Billion but will not be tackled in details in this book as it covers the years 2007-2013 and the programme's sub-categories, budget and topics are not yet available.

---

13 *Council Resolution concerning FP for Community RTD&D and FP1 (1984-1987)*; OJEC, 4/8/1983; *Decision of the Council concerning the FP2 for Community RTD activities (1987-1991)*, OJEC, 24/10/1987; *Decision of the Council concerning the FP3 for Community RTD activities (1990-1994)*, OJEC, 8/5/1990; *Decision of the EP and of the Council concerning the FP4 of the EC for RTD&D activities (1994-1998)*, OJEC, 18/5/1994; *Decision of the EP and of the Council concerning the FP5 of the EC for RTD&D activities (1998-2002)*, OJEC, 1/2/1999; *Decision of the EP and of the Council concerning the FP6 of the EC for RTD&D, contributing to the creation of the ERA and to innovation (2002-2006)*, OJEC, 29/8/2002.

Figure II.3: Community RTD budget from FP1 to FP7 (in € Billion)

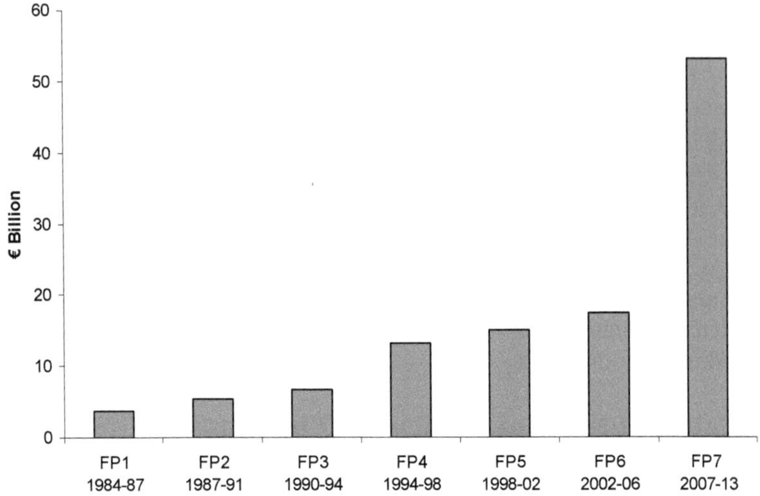

If the formal RTD Framework programmes budgets seem to have largely increased from FP1 to FP6, it is worthwhile mentioning that these figures do not reflect the whole EU research efforts in the considered period. From the middle of the 1980s, in addition to the ECU 3.8 Billion of FP1, it should be added, for example, the ESPRIT programme having a budget of ECU 750 Million for the 1984-1989 period[14] or the Joint Research Activities having a budget of ECU 700 Million for the 1984-1987 period[15] or the plan to stimulate European scientific and technical exchanges (mobility of researchers) having a budget of ECU 60 Million for the 1985-1988 period[16].

In the same way, to the ECU 5.4 Billion of FP2 and ECU 6.6 Billion of FP3, it should be added during this period, for example, the ECU 360 Million for demonstration projects in the energy field (1986-1989)[17], the ECU 75 Million for environment – air, soil, water quality, climatology, natural risks – (1986-1990)[18], the ECU 70 Million dedi-

---

14 *Council decision on a European strategic programme for research and development in information technologies (ESPRIT)*, OJEC, L 67, 9/3/1984.

15 *Council decision on a programme of research activities to be executed by the Joint Research Centre for the EAEC and EEC (1984-1987)*, OJEC, L 3, 5/1/1984.

16 *Council decision on a plan to stimulate European cooperation and exchange in the scientific and technical field*, OJEC, L 83, 25/3/1985.

17 *Council resolution on the promotion, by financial support, of demonstration projects and industrial projects in the energy field*, OJEC, L 350, 27/12/1985.

18 *Council decision on the pluriannual programmes of research in the field of environment*, OJEC, L 159, 14/6/1986.

cated to research activities in the materials programme including EURAM (1986-1989)[19], the Joint Research Centre activities budget of ECU 864 Million for the 1992-1994 period[20].

Concretely, only from FP4, i.e. the mid of the 1990s, the EU research activities – including Joint Research Centre and demonstration initiatives – have been included in the "RTD Framework Programme". Today, the majority of the EU RTD&D actions are therefore assembled in one single instrument – the RTD Framework Programme.

RTD is also part of the Community Structural Funds. From the beginning of the 1990s, about € 1 Billion per year are dedicated to RTD in the less favoured regions of Europe[21]. Among the main initiatives essentially coming from the European Regional Development Fund (ERDF) and its Article 10, and the European Social Fund (ESF) and its Article 6, it could be reminded the "Science and Technology for Regional Innovation and Development in Europe" (STRIDE), the "Development of Community dimension for vocational training and employment promotion measures" (EUROFORM), the Regional Innovation Strategies, and the Community initiatives concerning border areas (INTERREG).

In percentage of total Community RTD expenditures, the research budget remained relatively stable from the mid-1980s with a proportion oscillating between 3 and 4% of total Community expenditure (the EU budget was around € 100 Billion in 2004)[22]. With the 2000 Lisbon strategy, the 2002 RTD Barcelona targets and the 2007-2013 financial perspectives, the RTD expenditures have largely increased these last few years and could achieve up to around 8% of EU budget in 2013.

---

19 *Council decision on a research programme in the field of materials (raw materials and advanced materials)*, OJEC, L 159, 14/6/1986.

20 *Council resolution concerning the activities to be undertaken by the Joint Research Centre*, OJEC, C 118, 19/5/1992.

21 European Commission, *Communication on cohesion and research policy: synergies between research and technology development policy and cohesion policy*, COM(93)203 and European Commission, *Communication on the regional dimension of the European Research Area*, COM(2001)549.

22 European Commission, *Second European Report on S&T indicators 1997*, OPEC, Luxembourg, 1997 and own estimates.

Figure II.4: Trends in RTD expenditure as percentage of total Community expenditure (in percentage)

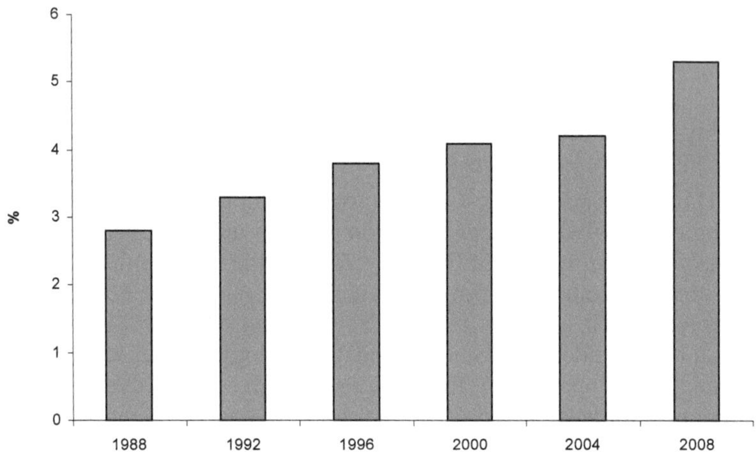

European Union energy RTD budgets

From the first to the third framework programme, the part of energy RTD has been largely reduced in comparison to other technologies and fields of research (see Figure II.5)[23].

In the latter phases (from FP3 period onward), the budget remained stable in terms of EU RTD percentage and in total value to around € 2 Billion including nuclear RTD (fission and fusion) and non nuclear RTD and demonstration (economic research in energy, fossil fuels, rational use of energy, renewable energy sources, fuel cells, hydrogen and storage). If energy was a key priority after the first oil crisis, new materials and life sciences acquired approximately the same importance in the 1980s and 1990s. From FP2, information and communication technologies RTD budget largely overstep the energy RTD.

The European Research Area launched in 2000 lays down the types of actions and instruments for the achievement of its objectives[24]. Under current practices, the research projects are selected by the Commission based on calls for proposals (and not from call for tenders) which leave a great degree of flexibility to researchers. The European Research Area

---

23 European Commission, *Research thematic priorities*, DG Research, Brussels, 2002.

24 European Commission, *Towards a European Research Area*, *op. cit.*; European Commission, *Making a reality of the European Research Area: guidelines for EU research activities*, *op. cit.*

– and its financial tool, the FP6 – provides specific types and forms of research management:

- Integrated Projects (IP) and Specific Targeted Research Projects (STREP) which are aimed at generating, demonstrating and validating knowledge through research and development. These projects are focused on some priority themes to be executed by research consortia.
- Networks of Excellence (NoE) aiming at bringing together the public or private scientific capacity of excellence for the implementation of long term commonly agreed activities. NoE are aimed at the durable integration of the participants' activities/capacities.
- Coordination Actions (CA) and Specific Support Actions (SSA) are aimed at supporting collaboration and coordination, and other activities (studies, conferences, etc.).
- Cooperation among national programmes with the aim of their mutual opening and their co-ordinated execution (art. 169).

In addition, it envisages support for innovation and start-ups in SMEs, for research infrastructure, for new and emerging scientific and technological issues, for development of human potential as well as for bringing closer research and society and for "Scientific Support to Policies" actions.

In the on-going FP7 – not detailed in this book – new instruments like Joint Technology Initiatives, ERA-NET and Joint Programming are new ways of pooling together the private and Member States research efforts at the European-level.

Figure II.5: Thematic priorities within the six EU RTD Framework Programmes (in percentage)

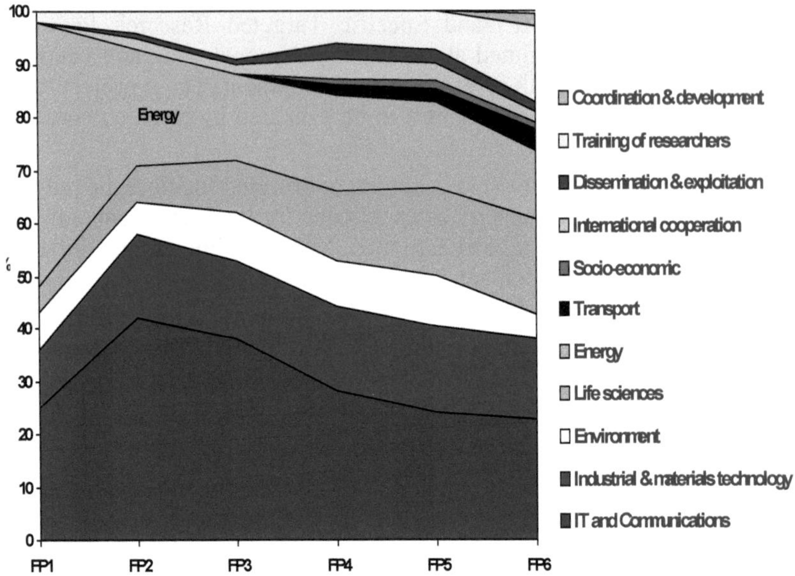

This EU energy RTD tendency is comparable to the international trend as demonstrated in the OECD Countries (IEA figures)[25]. IEA government energy RTDs budget peaked in 1981 at $ 15 Billion ($ and € were approximately equivalent in 2002), but then declined to about $ 9 Billion in 1987. From 1987 to 2002, funding was relatively stable, averaging about $ 9 Billion from 1987 to 1991 and $ 7.5 to 8 Billion in the 1990s. Total energy RTD&D expenditures in 2002 were just under $ 8 Billion.

25 International Energy Agency, *Statistics*, Paris, 2002 and own estimates for 2004.

Figure II.6: Government energy RTD budgets in IEA countries (in \$ Billion$_{2002}$)

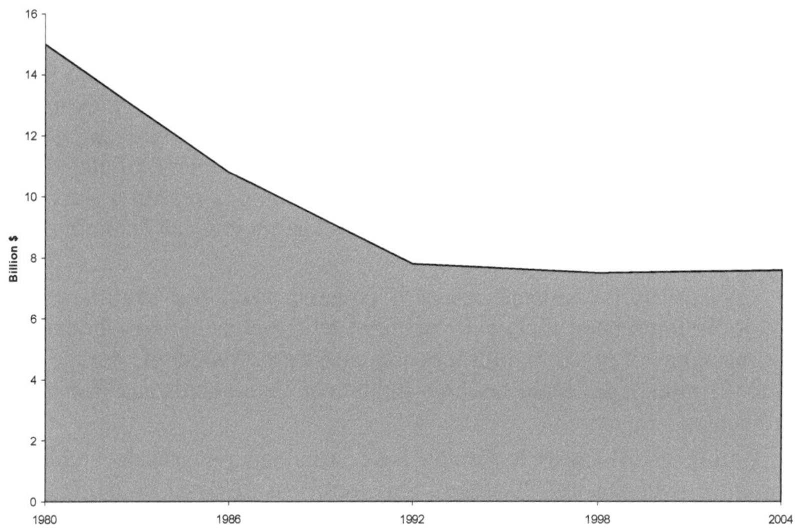

European Union nuclear and non nuclear energy RTD

Within the EAEC Treaty (Euratom) in 1957, the promotion of nuclear research was seen as one of its three objectives. The other two were securing the peaceful use of nuclear fissile material and know-how (safeguards), and the creation of a strong nuclear industry within the six Member States at that time. Until today, the provisions of the Treaty have never been changed, even if the number of the Member States has more than quadrupled from six in 1957 to twenty seven today.

Two tools have been used for promoting nuclear energy:

- Direct support to nuclear research provided through the Commission's Joint Research Centre, established by the Euratom Treaty;
- Indirect support to nuclear research provided through Community participation in shared-cost activities at European level.

Nuclear fusion was, from the beginning, reactor-oriented, with the objective of demonstrating the scientific and technological feasibility of fusion and aiming – in the long-term – at the joint construction of environmentally benign, safe and economic fusion reactors for electricity generation. The EU activities were always programme-oriented and institutionalised through long-term bilateral and multilateral contracts with the associated laboratories and institutions in the Member States and Switzerland.

As an integrated programme, Fusion activities at EC level represent around 40% of European effort in this field, i.e. more than the average 25% of the EC non nuclear activities.

Controlled thermonuclear fusion kept a budget of about € 750-850 Million in FP4, FP5 and FP6. The fully integrated European fusion RTD programme, which embraces all fusion energy research activities in the Member States and Switzerland, includes the continued operation of the Joint European Torus (JET) facilities, the improvement of the basic concepts of fusion devices and long-term technology research activities and the construction of the International Thermonuclear Experimental Reactor (ITER).

Nuclear fission related research covered from the beginning the whole fission related fuel cycle: mainly fuel enrichment and fabrication, various types of reactors, reprocessing, and waste treatment, storage and final disposal. Additional research fields were Safeguards and Radiation Protection.

Resources allocated to fission have remained remarkably stable in current € over the last 20 years. This trend has affected both JRC and shared cost activities:

- the nominal nuclear EC expenditure through the JRC has been kept roughly constant at € 300 Million/FP, albeit with a certain tendency to decrease;
- the same holds in a similar way for the expenditure on indirect support to research on fission related topics, with an average value of about € 190 Million/FP.

Within this overall apparent and nominal stability, significant changes occurred in the research portfolio reflecting both the progress of scientific activity in that field and the political priorities of the time:

- Certain stability has prevailed in areas such as safety, radiation protection and safeguards, which also correspond to missions entrusted by the Treaty to the EC. This is especially true for safeguards research where the JRC plays a prominent role, and for radiation protection research providing the required scientific basis for basic safety standards of the EU;
- Some research activities have been discontinued: decommissioning research for instance has produced mature and well-mastered technologies and approaches, which has lead to reduce support in this field to dissemination of knowledge;
- Other areas have been more subject to the influence of changing political needs and interest. This is the case for waste management which has gained increased attention and extended since 1994 (FP4) to innovative topics such as partitioning and transmutation or new

concepts aiming at minimising nuclear waste and/or optimising the use of the energy content of the nuclear fuels.

Innovation in nuclear fission implies tough challenges of world-wide interest, extending over three time horizons: today, 2010 and 2040. Three generations of nuclear power technologies, called "II", "III" and "IV", are distinguished over this period[26]. To each generation, a couple of additional challenges are associated:

- Generation II (today): security of supply and environmental compatibility;
- Generation III (around 2010): enhanced safety and competitiveness (economics);
- Generation IV (around 2040): cogeneration of heat and power, and full recycling.

The FP6 research programme extended over the period 2002-2006 and consisted of the EC part (non nuclear research: € 16270 Million) and the Euratom part (nuclear research: € 1230 Million, broken down into 750 for *fusion* and 480 for *fission*, i.e. 190 Million for *indirect actions* organised by DG Research and 290 for *direct actions* carried out by DG JRC). The total FP6 research budget (EC and Euratom programmes, direct and indirect actions) was € 17500 Million. On an annual basis, this EU budget corresponded to circa 5% of the public civilian research budget of the EU Member States and 4% of the EU budget.

[26] G. VAN GOETHEM *et al.*, *Euratom research and training in reactor systems and fuel cycles: EU response to some of the main S/T challenges in nuclear fission*, FISA-2006 Conference "EU Research and Training in Reactor Systems", Luxembourg, 13-16 March 2006.

Figure II.7: Energy research in the EU RTD Framework Programmes (in nominal € Million)

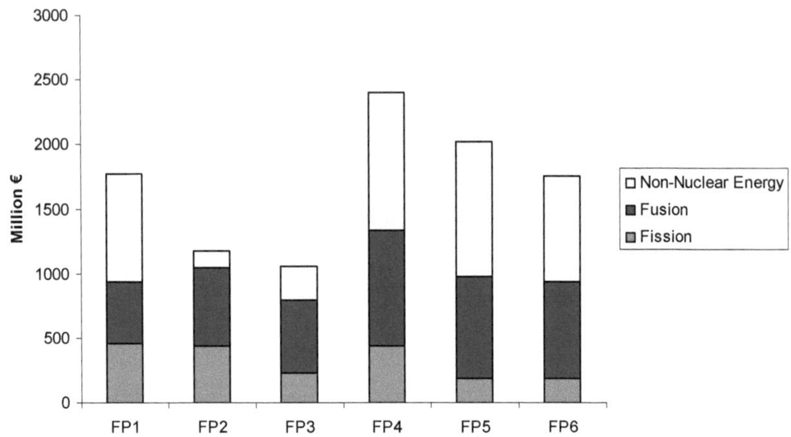

Concerning *Non nuclear* energy, it should be noted that the great difference between FP3 and FP4 is relatively artificial and explained by the fact that in FP3, NNE demonstration activities accounting for € 700 Million (Thermie Programme 1990-1994) were not part of the FP (and therefore not indicated in this Figure). In the same way, it should be reminded that in the 1980s, the NNE demonstration was not covered by the FP (and therefore, again, not indicated in this table) but by Council resolutions[27]. These resolutions concern: alternative energy sources, energy saving and substitution of hydrocarbons (€ 70 Million for 1983-1985 period); hydrocarbons (€ 140 Million) and demonstration projects and industrial pilot projects in the energy field (€ 360 Million) for the 1986-1989 period.

Concerning nuclear fission, the large variations between FP3 and the other Framework Programme are affected by changes in the EU accounting systems. A large part of the research activities in nuclear fission is directly led by the EU through its JRC – Joint Research Centre (mostly Petten, Karlsruhe and Ispra sites). In the Figure II.7, these JRC activities are only included in FP4. For the other FP, the nuclear fission JRC activities can be estimated to around € 350 Million in FP1, 310 in FP3, 280 in FP5 and 250 in FP6.

[27] *Council resolutions 1972/83*; *Council resolution on a programme of support for technological development in the hydrocarbons sector*, OJEC, L 350, 27/12/1985; *Council resolution on the promotion, by financial support, of demonstration projects and industrial projects in the energy field*, OJEC, L 350, 27/12/1985.

It should also be mentioned that a part of coal research (coal production, preparation and transport, coal combustion, and coal conversion) is funded by the ECSC. In the 1990s, the average ECSC financial support to coal research has been around € 30 Million per year (with a peak of € 50 Million in 1993 and with the lowest funding of € 18 Million in 1995).

Figure II.8 shows the total EU energy RTD&D effort in the Framework Programme periods including not only the FP budgets but also the other European above-mentioned efforts in energy research.

Figure II.8: EU energy RTD&D budget including FP budget, demonstration for non-nuclear energy and JRC for nuclear fission (in nominal € Million)

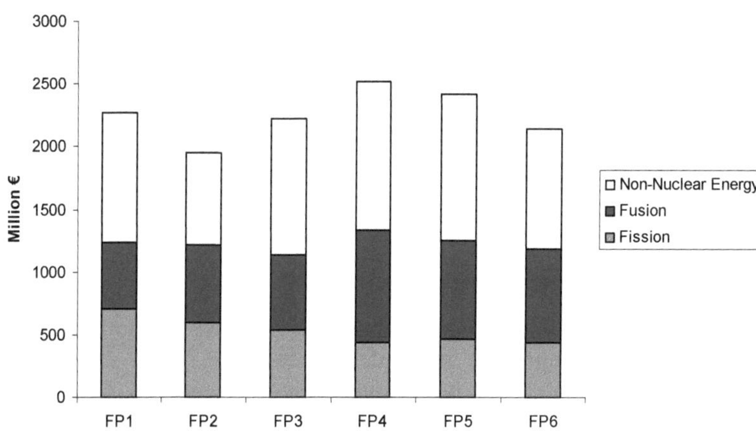

Taking into account these figures in real terms rather than in nominal terms and considering the successive EU enlargements (Portugal and Spain in 1986; Austria, Finland and Sweden in 1995; and Cyprus, Czech Republic, Estonia, Hungary, Latvia, Lithuania, Malta, Poland, Slovakia and Slovenia in 2004), the EU energy RTD&D per inhabitant was € 3 per year in the middle of the 1980s, € 2 in the 1990s and € 1 in the middle of the decade of the 2000 decade[28].

[28] In real terms (euro of 2004, i.e. middle of FP6), with the average conversion factor of 66 for FP1, 70 for FP2, 78 for FP3, 85 for FP4 and 93 for FP5, the figures were the following: 3,400 M€ for FP1, 2,900 M€ for FP2 and FP3, 3,000 M€ for FP4, 2,600 for FP5 and 2,250 for FP6. The EU population went from less than 300 millions people in FP1 (before Spanish and Portuguese accession) to 450 millions after the 2004 enlargement to Eastern and Southern Europe. *De facto*, taking into account the energy RTD&D FP figures adjusted for differences in the price level from 1984 to 2006 and considering the successive EU enlargements in this period, the EU energy

The general tendency to see a reduction of the energy RTD, is also demonstrated at the international level taking into consideration the IEA figures.

Figure II.9: Government energy RTD budgets in IEA countries (in $ Billion$_{2002}$)

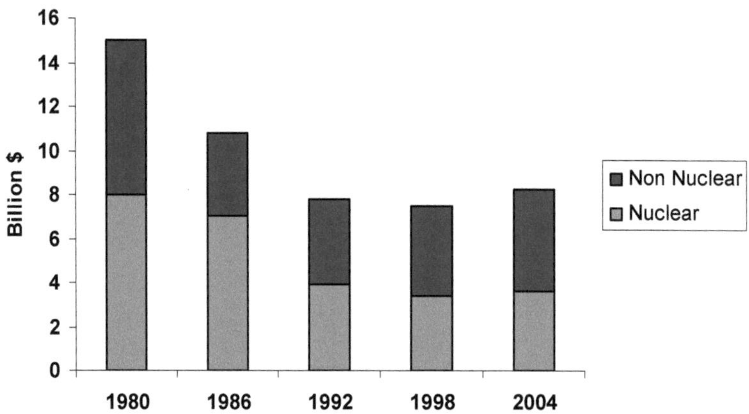

European Union energy economic research

The 1973 first oil crisis contributed to establish the political consensus for the Community's non nuclear energy RTD. It started with a long-term demonstration programme that concentrated essentially on *hydrocarbons* exploration and production, especially in order to develop cost effective technologies for tapping the "marginal" resources of the North Sea.

Progressively the portfolio was enlarged to cover also solid fossil fuels, rational use of energy and renewable energy sources[29]. A research programme dealing with long-term approaches was set up in 1975, along the same three broad priorities. The subject of systems analysis (socio-economic research in energy dealing essentially with models, scenarios and externalities) has been a permanent feature of these research activities, with a view to provide support to the decision-makers in energy policy and energy RTD matters. The objective of the European Commission was to develop the most effective analytical system to

RTD&D spending per inhabitant were € 3 per year in the middle of the eighties, € 2 in the 1990s and € 1 in the middle of the 2000 decade.

29 The ECSC Treaty had provided for an appropriate framework for RTD in coal long before the first decisions were made on other fossil fuels.

investigate the links between energy and the rest of the economy, long-term energy demand and the interactions of energy demand and supply[30].

Before 1994, there were two separate legislations: one establishing research programmes and another one for demonstration programmes. Since 1994, research and demonstration have been integrated within the same legal instrument (RTD&D Framework Programme). Originally, demonstration received much more financial support than research; after the first oil crisis, the objective was to find sustainable alternatives to oil and decrease Europe's dependence on imports. From FP4 to FP6, there is a half-and-half repartition of the budget between research and demonstration and since the mid of the 1980's, EC support to hydrocarbons RTD has been gradually decreasing in a relative way[31].

The FP budgets decided by the Council and – since the implementation of the Amsterdam Treaty – the European Parliament cover the whole programmes such as "energy" and do not expect a sub-division, for example between fossil fuels research or renewable research. The only way to get a clear picture of the research expenditures is to list all the "projects" funded under the various FP. As indicated in the OECD Frascati manual related to the R&D expenditures and to have a coherent picture in the various FP, only the purely research projects, i.e. not including the demonstration projects, are indicated in Table II.1[32].

---

30 R. RUBIN, E. ROMBERG, *Energy models for the European Community*, in A. STRUB (ed.), IPC Science and Technology Press, 1979.

31 European Commission, *JOULE I – catalogue of contracts*, programme 1989-1992, OPEC, Luxembourg, 1993; European Commission, *JOULE II – project synopses*, programme 1990-1994, OPEC, Luxembourg, 1994; European Commission, *JOULE III – project synopses*, programme 1994-1998, OPEC, Luxembourg, 1999; European Commission, *Socio-economic projects in energy and environment*, OPEC, Luxembourg, 2001; European Commission, *Energy – Economic, social and policy-oriented research in Europe*, OPEC, Luxembourg, 2007.

32 "Research and experimental development (R&D) comprise creative work undertaken on a systematic basis in order to increase the stock of knowledge, including knowledge of man, culture and society, and the use of this stock of knowledge to devise new applications; R&D is a term covering three activities: basic research, applied research and experimental development. Basic research is experimental or theoretical work undertaken primarily to acquire new knowledge of the underlying foundations of phenomena and observable facts, without any particular application or use in view. Applied research is also original investigation undertaken in order to acquire new knowledge. It is, however, directed primarily towards a specific practical aim or objective. Experimental development is systematic work, drawing on existing knowledge gained from research and or practical experience, that is directed towards producing new materials, products and devices; to installing new processes, systems and services; or to improving substantially those already produced or installed. The basic criterion for distinguishing R&D from related activities is the presence in R&D of an appreciable element of novelty and the resolution of scientific and/or technological

Concretely, the following data should be referred to the Figure II.7. For FP2 and FP3, the figures in the Table II.1 are identical to those in Figure II.7. For FP4, FP5 and FP6, the data in Table II.1 only accounts for half of those indicated in Figure II.7. In fact, from 1994 (FP4), demonstration projects – which accounts for fifty per cent of the non nuclear energy programme – have been included in the Framework Programmes. In order to cover the different thematic priorities in the various FP, the following *ad hoc* classification has been established for FP2 to FP6:

- Economic research in energy, including model developments, scenarios and external costs quantification;
- Rational use of energy/Energy efficiency including in buildings and industry, for air conditioning, for lighting control and for heat pumps;
- Fossil fuels, including deep reservoirs
- CO_2 capture and storage;
- Fuel cells, including storage, hydrogen and electricity;
- Renewables, including wind, photovoltaics, biomass/waste, geothermal, solar and ocean technologies.

This classification has been done by the author on a project by project basis from FP2, i.e. from 1987 to the end of FP6, i.e. to 2006.

uncertainty, that is, when the solution to a problem is not readily apparent to someone familiar with the basic stock of commonly used knowledge and techniques in the area concerned". Definition from: OECD, *Frascati Manual*, Paris, 2002.

Table II.1: European Union non-nuclear energy research budget from 1987 to 2006 without demonstration and without JRC (in nominal ECU or €)

FP2 (1987-1991)	Budget	Percentage
Economic research in energy	4,622,000	4%
Rational use of Energy/Energy efficiency	15,969,000	13%
Fossil Fuels	44,682,000	38%
CO_2 capture and storage	0	0%
Fuel Cells/Hydrogen/Storage	12,537,000	10%
Wind	9,566,000	8%
Photovoltaics	13,023,.000	11%
Biomass	10,931,000	9%
Geothermal	7,931,000	7%
Solar	0	0%
Ocean	520,000	0%
Hybrid RES/Integration/RES in Buildings	0	0%
Total	**119,781,000**	100%

FP3 (1990-1994)	Budget	Percentage
Economic research in energy	12,774,000	5%
Rational Use of Energy/Energy efficiency	30,867,000	13%
Fossil Fuels	65,517,000	27%
CO_2 capture and storage	800,000	0%
Fuel Cells/Hydrogen/Storage	32,577,000	13%
Wind	24,753,000	10%
Photovoltaics	29,145,000	12%
Biomass	10,017,000	4%
Geothermal	13,225,000	5%
Solar	692,000	0%
Ocean	3,920,000	2%
Hybrid RES/Integration/RES in Buildings	18,832,000	8%
Total	**243,119,000**	100%

FP4 (1994-1998)	Budget	Percentage
Economic research in energy	24,466,000	6%
Rational use of Energy/Energy efficiency	53,913,000	13%
Fossil Fuels	47,325,000	11%
CO_2 capture and storage	0	0%
Fuel Cells/Hydrogen/Storage	73,470.000	17%
Wind	43,797,000	10%
Photovoltaics	78,260,000	18%
Biomass	49,181,000	12%
Geothermal	8,500,000	2%
Solar	11,637,000	3%
Ocean	5,483,000	1%
Hybrid RES/Integration/RES in Buildings	28,840,000	7%
Total	**424,872,000**	100%

FP5 (1998-2002)	**Budget**	**Percentage**
Economic research in energy	11,976,000	2%
Rational Use of Energy/Energy efficiency	9,327,000	2%
Fossil Fuels	87,388,000	18%
CO_2 capture and storage	18,032,000	4%
Fuel Cells/Hydrogen/Storage	124,336,000	26%
Wind	29,208,000	6%
Photovoltaics	60,969,000	13%
Biomass	53,337,000	11%
Geothermal	5,000,000	1%
Solar	9,293,000	2%
Ocean	4,475,000	1%
Hybrid RES/Distributed generation/ RES in Buildings/Electricity	73,354,000	15%
Total	**486,695,000**	100%

FP6 (2002-2006)	**Budget**	**Percentage**
Economic research in energy	20,490,000	4%
Rational Use of Energy/Energy efficiency	0	0%
Fossil Fuels	0	0%
CO_2 capture and storage	73,454,000	15%
Fuel Cells/Hydrogen/Storage	145,127,000	30%
Wind	15,320,000	3%
Photovoltaics	71,784,000	15%
Biomass	77,015,000	16%
Geothermal	12,500,000	3%
Solar	10,942,.000	2%
Ocean	12,580,000	3%
Hybrid RES/Distributed generation/Electricity	43,533,000	9%
Total	**482,745,000**	100%

Economic research in energy also called *Strategic energy research* or *Socio-economic research* in support to policies consistently received between 2 and 6% of the EU funding for non-nuclear energy research and was essential in providing the scientific basis of EU policy-making in energy-related fields.

Leaving aside demonstration, EU *fossil fuels* research decreased significantly in relative terms from 38% (€ 45 Million in FP2) to 0% in FP6. CO_2 capture and storage technologies started to be funded since 1998 and accounted for 15% (€ 73 Million) in FP6.

EU research in *renewable* energies increased from 35% in FP2 (€ 42 Million) to 41% in FP3, 47% in FP4, 49% in FP5 and 51% (€ 241 Million) in FP6. Photovoltaics was the technology that consistently received the highest funding except in FP6. Biomass was the second technology in terms of budget, except in recent years when it took over first place. Wind was important in the 1990s. Geothermal generally receives around € 2-3 Million per year and ocean and solar together received around 4% of the EU non-nuclear research budget from the mid-1990s. Statistics for the last category, "hybrid renewable, integration, renewables in buildings, distributed generation and electricity", are always difficult to obtain, as this category was generally oriented towards renewables but also connected to micro-generation and to the electrical grid. The figures do not take into consideration "demonstration projects" which account for about half of the budget since more than ten years.

Research on *fuel cells, hydrogen and storage*, grew steadily in importance from € 13 Million in FP2, to € 33 Million in FP3, to € 73 Million in FP4, € 124 Million in FP5 and € 145 Million in FP6. *Hydrogen* research was mentioned as one such key area of the first non-nuclear energy programme (1975-1979) and it attracted 20% of the overall budget. Later, interest in this area waned and it was mainly covered under fuel cell research. In FP5 and especially in FP6, hydrogen attracted renewed interest. The figures mentioned in Table 1 do not include the demonstration activities and the "Transport" programme activities the have also considerable budgets for these technologies.

Research on *Energy efficiency/Rational use of energy*, which embraces various sectors and technologies, accounted for € 16 Million in FP2, € 31 Million in FP3, € 54 Million in FP4, € 9 Million in FP5 and zero in FP6. Again, these figures do not include several demonstration activities that have been considered recently under the CONCERTO and CIVITAS programmes.

Figure II.10: EU non-nuclear energy research – without demonstration and JRC (in nominal € Million)

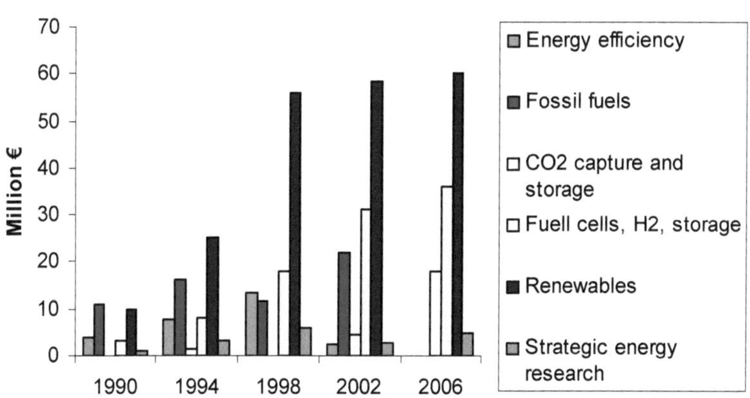

Table II.2: European Union projects in the field of economic research in energy (also called Strategic energy research or Socio-economic research) from 1989 to 2006 – FP2-FP6

Central applications of the Commission's models: MEDEE, EFOM, MIDAS, HERMES	1989
Models for energy and environment: improvement and case studies	1989
Data processing data banks	1989
Development and applications of the HERMES model (1)	1989
Development and applications of the HERMES model (2)	1989
Energy and environment project (1)	1989
Energy and environment project (2)	1989
International and long term energy models	1989
CO_2 study and fuel cycle costs project	1990
Proposal for work on EC/US joint study on fuel cycle costs	1991
External costs of fuel cycles	1991
External costs of fuel cycles	1991
World dimension and uncertainty	1991
Impacts of energy tax on the European economies	1991
Scientific advice for the EC/US joint study on fuel cycle costs	1991
MURE, COST, Biomass project completion	1991
Finalisation of the CO_2 study/CRASH Programme	1991
Compétitivité et politique d'environnement	1992
Energy – Technology – Environment – Competitiveness: Analysis and modelling of policy strategies	1992
Follow up of the EC/US study on the external costs of fuel cycles	1992
Social cost of rational use of energy	1992
Renewable energy in EC/US external costs of fuel cycle project	1992
A computable price-driven equilibrium model of the energy system and markets for Europe-PRIMES	1992
World long term energy systems: a model and global outlook for European strategies	1992
A general energy-environmental-economy model for Europe – E3ME	1992
National implementation of EC/US fuel cycle costs framework, The Netherlands case	1992
External costs of fuel cycles: implementation of the accounting framework in France	1992
The general equilibrium modelling project for energy-economy-environment	1992
Modelling economy-energy-environment linkages: A European panel approach	1992
SOLFEGE – TEG – SAFIRE – SITRE models	1992
External costs of fuel cycles – EC/US project phase II	1992
Renewable energy strategies	1992
Developing decision-making support tools for the utilisation of renewable energies in integrated systems at the local level	1992
Combined multiple renewable energy source system simulator facility	1992
Formulating a strategy for energy R&D: an advisory panel on methodologies	1993
Energy-Environment Modelling in Central and Eastern European Countries	1993
Energy R&D strategies for a sustainable future	1993
Benefits and costs of policy instruments for energy conservation	1993
Models for the study of transport energy and environment policies. Transferable models for transport and energy policy testing (TRENEN)	1993
External cost of fuel cycle. National implementation for Spain, Portugal and hydropower	1993
Integrated transport effects model system on urban scale (ITEMS)	1993
Dissemination of external costs accounting framework	1993
Green accounting: comparison of monetary valuation and 'EXTASY' method	1993

Sustainable, alternative, fluent energy-based-Mediterranean economic development	1993
Climate technology strategy within competitive energy markets: towards new and sustainable growth	1995
The implementation of new energy technologies: non technical barriers and policy responses	1995
Consumer user patterns of electrical vehicles. Application feasibility technical options, mobility policy and environmental assessment of electric cars	1995
Method for integrated evaluation of benefits, costs and effects of programmes for promoting energy conservation	1995
Interdisciplinary analysis of successful implementation of energy efficiency in the industrial, commercial and service sector (INTERSEE)	1995
Social and organizational issues in the adoption of advanced energy technologies in manufacturing	1995
Maintenance, improvement, extension and application of the ExternE accounting framework	1995
External costs of transport in ExternE	1995
The national implementation in the EU of the ExternE accounting framework	1995
Completion and extension of E3ME	1995
Synergies between European and national strategies for energy	1995
Scenario-based framework for modelling transport technology deployment: energy-environment decision support	1996
Fuelwood in Europe for environment and development strategy	1995
Total costs and benefits of biomass in selected regions of the European Union	1995
Biomass application in utility system	1995
EURE utility wind farm project	1995
Integrated spatial potential initiative for renewable energy in Europe (INSPIRE)	1996
The value of renewable electricity	1996
Financial analysis of the integration of renewables in electricity supply (FAIRE)	1996
Financing of renewable energy systems (FIRE)	1996
European scenarios on transport-energy-environment for metropolitan areas	1995
Scenario-based framework for modelling transport technology deployment: energy-environment decision support	1996
Mediterranean energy markets appraisal: trend and prospects for new technologies	1997
Effective policy instruments for energy efficiency in residential space heating – an international empirical analysis	1997
An expert system to integrate renewable energy sources in European energy supply system	1997
Voluntary agreements – implementation and efficiency	1997
Reducing barriers to energy efficiency in private and public organisations	1997
Non technical aspects of demand side management in European electricity markets	1997
Energy technology dynamics and advanced energy system modelling	1997
External costs of energy conversion – improvement of the ExternE methodology and assessment of energy-related transport externalities	1997
European emission mitigation policy and technological evolution: economic evaluation with the GEM-E3-EG model	1997
Industrial benefits and costs of greenhouse gases abatement strategies: applications of E3ME	1997
External costs discussion group	1998
Priority setting initiative: synergies between European and national energy RTD priorities	1997
Feasibility study on biofuel chains implementation in a Mediterranean country	1996
Electricity tariffs and embedded renewable generation	1998
Integration of renewable energies in the Southern Mediterranean region	1998

The value of renewable electricity – part 2 (REVALUE 2)	1998
Renewable electricity and liberalised markets	1998
Renewables burden sharing assessment tool (REBUS)	2000
The role of innovation and policy design in energy and environment for a sustainable growth in Europe (TECH-GEM-E3)	2000
System analysis for progress and innovation in energy technologies (SAPIENT)	2000
Scenarios and strategies for the implementation of the clean development mechanism of the Kyoto Protocol in the Mediterranean region (CDMED)	2000
Joint implementation for international emissions reduction through electricity companies in the European Union and in the central and eastern European countries (JOINT)	2000
Government regulatory energy measures impact and diffusion speed appraisal method (GRIDS)	2000
Poles world energy/technology outlook to 2030 (WETO 2030)	2000
Experience curve: a tool for energy policy programmes assessment (EXTOOL)	2000
New elements for the assessment of external costs from energy technologies (NEWEXT)	2000
European network for energy economics research (ENER)	2000
Scenarios for integration of renewables in a European cities network (SIREN)	2001
Establishment of a European energy data exchange network (EDEN)	2001
Very long term energy environment model (VLEEM)	2001
Assessing climate response options: policy simulations – insights from using national and international models (ACROPOLIS)	2001
Baselines for accession states in Europe (BASE)	2001
Enhancing the government regulatory energy measures impact and diffusion speed appraisal method (E-GRIDS)	2001
Experience curve analysis of photovoltaic energy systems and components (PHOTEX)	2001
Environmental and ecological life cycle inventories for present and future power systems in Europe (ECLIPSE)	2001
New econometric model for environment and strategies implementation and sustainable development/endogenous technical change (NEMESIS/ETC)	2001
Policy and regulatory roadmaps for the integration of distributed generation and the development of sustainable electricity networks (SUSTELNET)	2001
Strategic assessment framework for the implementation of rational energy – local planning (SAFIRE-LP)	2002
Deriving optimal promotion for increasing the share of renewable electricity in a dynamic European electricity market (GREEN-X)	2002
Technology and social visions for Europe's energy future – a Europe wide Delphi study (EURENDEL)	2002
Very long term energy assessment model – phase 2: assessment of energy RTD strategies for sustainability (VLEEM-2)	2002
Systems analysis for progress and innovation in energy technologies for integrated assessment (SAPIENTIA)	2002
Electricity market liberalisation in Europe (EMELIE); market imperfections and its implications on the European economic and environmental situation.	2002
The dynamics of innovation and investment and its impact on policy design in energy and environment for a sustainable growth in Europe (DYN-GEM-E3)	2002
Dissemination and discussion of the ExternE methodology and results (DIEM)	2002
Research & development spending: a survey of R&D spending for renewable energies in EU countries (REDS)	2002
Promoting and financing clean development mechanism renewable energy projects in the Mediterranean region (CDMEDI)	2002
Accession countries conference – electric power supply strategy in the 21st century – the renewable energy sources option (EPS 2003 – RES)	2003
New energy externalities developments for sustainability (NEEDS)	2004

Local new energy technology implementation (LETIT)	2004
World energy technology outlook-2050 (WETO H2)	2004
Case study comparisons and development of energy models for integrated technology systems (CASCADE MINTS)	2004
Sustainable energy specific support assessment (SESSA)	2004
Dissemination of external costs of electricity supply-making electricity external costs known to policy-makers (MAXIMA)	2004
Energy corridor optimisation, for the European markets of gas, electricity and hydrogen (ENCOURAGED)	2005
European sustainable electricity; comprehensive analysis of the future European generation of electricity and security of supply (EUSUSTEL)	2005
Scientific reference system on new energy technologies, energy end-use efficiency and energy RTD (SRS NET&EEE)	2005
Future energy technologies for enlarged European Union (FET-EEU)	2005
Large-scale integration of RES – E and cogeneration into energy supplies in accession countries (CEERES)	2005
Renewable Energy in developing countries: Current situation, market potential and recommendations for a win-win-win for EU industry, the environment and local socio-economic development (RECIPES)	2005
Electricity roadmap in Europe (ERMINE)	2006
The potential of transferring and implementing sustainable energy technologies through the Clean Development Mechanism for the Kyoto Protocol (ENTTRANS)	2006
Acceptance of CO_2 capture, storage – economics, policy and technology (ACCSEPT)	2006
Modelling of energy technologies prospective in a general and partial equilibrium framework (MENGTECH)	2006
Cultural influences on renewable energy acceptance and tools for the development of communication strategies to promote acceptance among key actor groups (CREATE ACCEPTANCE)	2006
Cost assessment for sustainable energy systems (CASES)	2006
Macro-economic impact of high oil price (HOP!)	2006
Coordinating energy security of supply actions (CESSA)	2006
RTD&D to increase the share of renewables in emerging and developing countries with European technologies; policy assessment; stakeholders opinions, best practices and recommendations (RTD4EDC)	2006

CHAPTER III

Scientific Controversies on Energy Foresight

Summary

The main objectives of this chapter aim to illustrate the scientific debate on quantitative and qualitative tools dealing with energy futures and with the evaluation of impacts of policies and measures.

Since the oil shocks of the 1970s (security of supply issue) and the concerns about global warming in the 1980s (climate change issue), energy policy became an important strategic matter at national and international levels and the need for good tools of policy analysis arose.

If most European decision-makers can agree on general principles like the internalisation of external costs or the application of the precaution principle, several ways could be taken to achieve these goals (environmental taxation, policies and measures or emission quotas) and their expected economic impacts have to be assessed in a medium or long-term time horizon.

These last years, the debate around energy technologies to reduce greenhouse gas emissions has demonstrated that there is no one single answer. There are many technical solutions and choice depends on social preferences and political possibilities. To reveal these preferences, public (and private) decision-makers look at both Qualitative and Quantitative outlooks. Decision support tools must be able to provide this range of outputs.

For looking at the future energy supply and demand, for measuring future trends in the emission of greenhouse gases and for assessing impacts of alternative policies and measures, different Qualitative and Quantitative scientific methods have been developed.

Under the Qualitative methods, "Foresight" tools essentially based on human inputs are considered. As Quantitative tools, mathematical models are analysed. "Top-down" models are economic-based and "Bottom-up models are engineering-based. For long-term issues (more than 50 years), the Shell and IPCC scenarios are examined. When dealing with very long-term (2100) and potential breakthroughs, the Back-casting approach is used.

These Qualitative and/or Quantitative outlooks help make sense of unpredictable multiple futures under different framework conditions. They provide coherent pictures of futures developed within a framework of specific assumptions. Qualitative and/or Quantitative outlooks are less about predicting than preparation of "what if" a specific event occurs or a specific measure is taken.

In the energy sector, these outlooks help public organisations like the International Energy Agency, the US Department of Energy or the European Commission to master change and to adopt strategies that take into account the costs and benefits of such or such a policy.

This chapter aims firstly at highlighting the various energy outlooks and their categorisation. Between qualitative and quantitative approaches, between stakeholder (participatory) and expert (non-participatory) inputs, and between a wide socio-political scope and a focused techno-economic problem-solving scope, "Foresight" and "Modelling" activities are classified.

Secondly, the different world, international, European and national energy outlooks are distinguished according to their time-frame and the three kinds of expected futures theorised by Jacques Durand in terms of "Continuity", "Exploratory" and "Breakthroughs". The origin of the System analysis by the *Club of Rome* at the beginning of the 1970s or the prospects made by the *Global Scenario Group* are also described.

Comparisons with some major international studies are made. Similarities of results, including for example for the long-term (2050) and using different tools like MARKAL or POLES are emphasized.

Thirdly, this chapter highlights the characteristics of the Qualitative outlooks (action-oriented, open to alternative futures, participatory and multidisciplinary) largely based on the Foresight triangle: thinking the future, debating the future and shaping the future.

The main methods of Qualitative outlooks (Expert-based Delphi surveys, Critical technology studies, Participatory approaches like the European Awareness Scenario Workshop, and Technology Roadmaps are analysed. The main strengths and weaknesses from these qualitative exercises are identified.

Examples of European energy socio-economic research projects and results (see Chapter II) are given and some issues related to these methods are characterised. Beyond the energy field, European Commission foresight and long-term planning activities (cf. Cellule de Prospective, EU financial perspectives, RTD Framework Programmes) are also addressed.

This chapter shows that both Qualitative and Quantitative studies have been led at the EU-level in several domains. But in the energy-related fields (see Chapters III), the support to economic and energy models will be stronger. This is probably due to the fact that energy is in the same time sufficiently simple to be modelled (identifiable and largely interchangeable primary energy sources) and too complex to be studied without any formalization (interactions with the economic and environmental systems).

Methods and tools for looking at the futures and for evaluating policies

There are several – and sometimes overlapping – ways to look at the future (technological) developments, to assess their impacts and to try to manage risks and uncertainty. At the beginning of the 1920s, Frank H. Knight was one of the first to underline the relevance of risk and uncertainty in economic theory. According to Knight, risk refers to situations where the decision-maker can assign mathematical probabilities to the randomness that he faces. On the contrary, uncertainty refers to situations when this randomness cannot be expressed in terms of specific mathematical probabilities[1].

1 "Uncertainty must be taken in a sense radically distinct from the familiar notion of risk, from which it has never been properly separated. The term 'risk', as loosely used in everyday speech and in economic discussion, really covers two things which, functionally at least, in their causal relations to the phenomena of economic organization, are categorically different. The nature of this confusion (...) may be stated in a few words at this point. The essential fact is that 'risk' means in some cases a quantity susceptible of measurement, while at other times it is something distinctly not of this character; and there are far-reaching and crucial differences in the bearings of the phenomenon depending on which of the two is really present and operating. There are other ambiguities in the term 'risk' as well, which will be pointed out; but this is the most important. It will appear that a measurable uncertainty, or 'risk' proper, as we shall use the term, is so far different from an unmeasurable one that it is not in effect an uncertainty at all. We shall accordingly restrict the term 'uncertainty' to cases of the non-quantitative type". F.H. KNIGHT, *Risk, uncertainty, and profit*, Houghton Mifflin Company, Boston MA, 1921.
Fifteen years later, dealing with special-cause variation, John Maynard Keynes said: "By uncertain knowledge, I do not mean merely to distinguish what is known for certain from what is only probable. The game of roulette is not subject, in this sense, to uncertainty. The sense in which I am using the term is that in which the prospect of a European war is uncertain, or the price of copper and the rate of interest twenty years hence. About these matters there is no scientific basis on which to form any calculable probability whatever. We simply do not know". In its famous General Theory, Keynes concluded that classical economics rested on a fundamental error. It assumed, mistakenly, that the balance between supply and demand would ensure full employment. On the contrary, in Keynes's view, the economy was chronically unstable and subject to fluctuations, and supply and demand could well balance out at an equilibrium that did not deliver full employment. The reasons were inadequate investment and over-saving, both rooted in the psychology of uncertainty. J.M.

According to Kees van der Heijden, "Scenarios make uncertainty explicit"[2]. Foresight allows anticipating situation of crises. In his book *Pour un catastrophisme éclairé*, Jean-Pierre Dupuy hopes that "cet avenir, bien qu'inéluctable, n'ait pas lieu"[3].

A large part of the literature on methods and tools for looking at the futures and for evaluating policies has been published in France and in the USA. Among the modern pioneers (in the 1960s and the 1970s) in the field of Foresight ("*Prospectivistes*"), one notes the French scientists like Gaston Berger, Bertrand de Jouvenel, Pierre Wack, Pierre Massé and Pierre Piganiol. In the US research community important contributions have been made by Jay Forrester, Dennis Meadows and Mihajlo Mesarovic. Other significant work has been made by the British Alexander King, the German Eduard Pestel and the Italian Aurelio Peccei. Many of these researchers either were academics or connected with the Club of Rome, the Commissariat Général du Plan or with the Shell Group[4].

"Foresight", "Scenarios" and "Strategic Planning" are probably the most commonly used terminology to address the issues of looking at the futures and to evaluate policies. Jerome C. Glenn and Theodore J. Gordon distinguish more than fifty methods including Interviews, Cost-benefit analysis, Bibliometrics, Multicriteria decision analyses, Risk

---

KEYNES, *General theory of employment, interest and money*, Cambridge University Press, 1936.
In the context of environmental regulations in the United States, James Hammitt pointed out that "the existence of uncertainty does not prevent decision-making. Nor the absence of uncertainty necessarily resolve the difficulty of decision-making (…). Tools for analysing environmental decisions have proliferated with growing recognition of the complexity of the issues. In addition to general policy- analytic methods like Benefit-Cist and Cost-Effectiveness Analysis, the new tools include Risk Assessment, Comparative Risk Analysis, Risk Trade-off Analysis, Risk-Risk (or Health-Health) Analysis, and Life-Cycle Analysis". HAMMITT J., *Analytic methods for environmental regulations in the United States: promises, pitfalls and politics*, Int. J. Risk Assessment and Management, Vol. 1, No. 1/2, 2000.

2 K. VAN DER HEIJDEN, P. SCHUTTE, *Key questions for designing scenario applications, Scenario and Strategy Planning*, Vol. 1, No. 6, 2002.

3 J.P. DUPUY, *Pour un catastrophisme éclairé*, Seuil, Paris, 2002.

4 G. BERGER, *L'attitude prospective*, *op. cit.*; J. DURAND, *Prospective, discontinuité et instabilité*, Futuribles, *op. cit.*; M. GODET, *The art of scenarios and strategic planning: Tools and pitfalls*, *op. cit.*; P. SCHWARTZ, *The art of the long view: Planning for the future in an uncertain world*, Doubleday, New-York, 1991; P. WACK, *Scenarios: Uncharted waters ahead*, Harvard Business Review, Sept-Oct. 1985; T.J. CHERMACK, S.A. LYNHAM, W.E.A. RUANA, *A review of scenario planning literature*, Futures Research Quarterly, Vol. 17, No. 2, 2001; K. VAN DER HEIJDEN, R. BRADFIELD, G. BURT, G. CAIRNS, *The sixth sense*, Wiley, New York, 2002. H. MINTZBERG, *The Rise and fall of strategic planning*, The Free Press, New York, 1994.

analysis or Technology assessment[5]. According to Alan Porter, all these techniques fit into a field called "Technology futures analysis"[6]. Richard A. Slaughter argues that Future studies should become a genuine discipline[7].

In order to take into account the main economic (e.g. growth rate), social (e.g. demographic trends), environmental (e.g. CO_2 emissions) and technological variables considered in the energy field, we prefer to use the terminology "Qualitative and/or Quantitative outlook". The term "Outlook" is also closer to the one used in several energy studies performed by organisations like the International Energy Agency, the European Commission or the US Department of Energy[8]. Nevertheless, as it will be demonstrated latter, the quantitative component of these "Outlooks" largely predominate the qualitative one.

Under the Qualitative outlook, we consider essentially "Foresight" (including "Delphi" surveys, "Critical Technology" studies and "Technology roadmaps"). Under the Quantitative outlook, we focus on "Modelling" with the distinction between Top-down and Bottom-up models (and also Hybrid and Integrated models). As shown in Figure III.1, this categorisation is essentially based on the Qualitative or Quantitative approaches but also on the stakeholder (participatory) or expert (non-participatory) inputs, and on a wide socio-political scope or on a techno-economic problem-solving.

A Qualitative and/or Quantitative outlook exercise is generally launched when a country, an organisation or a firm has to face a specific challenge or has to take strategic decisions. This "Scenario planning" finds its roots in the military field. In particular, Herman Kahn, a defence analyst at the RAND Corporation in the USA in the late 1940s, started to describe stories of the many possible ways that nuclear weapons might be used by hostile nations. H. Kahn launched the concept of "thinking the unthinkable"[9]. In France, the philosopher Gaston Berger

5 T.J. GORDON, J.C. GLENN, *Futures research methodology*, Millennium project of the American Council for the United Nations University, July 2003.

6 A. PORTER, *Technology futures analysis: Toward integration of the field and new methods*, 2004.

7 R.A. SLAUGHTER, *Future studies as an intellectual and applied discipline*, American Behavioural Scientist, Vol. 42, No. 3, 1998.

8 International Energy Agency, *World Energy Outlook (WEO)*, Paris, 2003 and 2006; European Commission, *World energy, technology and climate policy outlook (WETO-2030)* and *World Energy Technology Outlook-2050 (WETO-H2)*, Luxembourg, 2003 and 2006; US Department of Energy (EIA), *International Energy Outlook (IEO)*, Washington, 2001 and 2006.

9 The RAND Corporation is an American think tank first formed to offer research and analysis to the US military. The organisation has since expanded to working with other governments and commercial organisations.

was among the first to speak about *Prospective* ("Foresight"). In this term, he involved the ideas of looking far and large, of in-depth analysis, of taking risks and of thinking to the man (*Voir loin, voir large, analyser en profondeur, prendre des risques et penser à l'homme*)[10]. Fifty years later, a national programme and graduate school in future studies will be launched in Finland (Turku School of Economics and Business Administration).

Qualitative and/or Quantitative outlooks have generally been developed by and for public purposes. Nevertheless, since the beginning of the 1970s, some private companies like Shell – under the thinking leadership of Pierre Wack – have also developed their own way of building scenarios that were widely recognised[11]. Such scenarios – close to a management innovation – have become a widespread consulting practice, popularized by writers like Peter Schwartz or Kees van der Heijden former officials in Shell's Group Planning department.

At the global level, various groups or think-tanks have as a major objective to look at the future. Among the most famous group, the *Club of Rome* takes always an interdisciplinary perspective (interaction of economic, political, social, cultural, ecological, scientific and technological problems) with a long-term approach indicating future trends. The Club of Rome published the well-known "Limits to Growth" in 1972 that modelled the consequences of a rapidly growing world population with finite resource supplies[12].

This exercise was based on the so-called "Systems Approach" or "Systems Dynamics" methodology proposed in 1971 by Jay W. Forrester in which causal networks are built in quantitative models (cf. *World2* model developed by J. Forrester and *World3* model that will be used by Dennis Meadows in "Limits to Growth")[13]. The controversies about the

---

10 G. BERGER, *L'attitude prospective*, *op. cit.*

11 P. WACK, *La planification par scénarios*, Futuribles No. 99, Paris, mai 1986; J. LESOURNE, Ch. STOFFAES, *La prospective stratégique d'entreprise – Concepts et études de cas*, *op. cit.* Several other companies like Motorola (see below) or Siemens have developed scenarios exercise. J. SCHARIOTH, M. HUBER, K. SCHULZ, M. PALLAS, *Horizons 2020 – A thought-provoking look at the future*, study commissioned by Siemens, TNS Infratest, Munich, 2005.

12 D.H. MEADOWS, D. MEADOWS, J. RANDERS, W. BEHRENS, *Limits to Growth*, *op. cit.* In the same time, in France, the philosopher and sociologist Edgard Morin published a paper on the beginning of a new ecological era (E. MORIN, *An I de l'ère écologique*, Nouvel Observateur, Paris, 1972). A second report of the Club of Rome will be published two years later (M. MESAROVIC, E. PESTEL, *Mankind at the Turning Point*, Club of Rome, 1974). In 1992, the book Beyond the Limits was published (D.H. MEADOWS, D.L. MEADOWS, J. RANDERS, *Beyond the limits – Global collapse or sustainable development*, Earthscan Publications, London, 1992).

13 J.W. FORRESTER, *World Dynamics*, Wright-Allen Press, Cambridge, 1971.

projections of this *Club of Rome* report (extinction of vital resources within 100 years) raised the attention on the complexity of the dynamic interactions among the energy, economic, social and natural systems.

In 1972, the IIASA (International Institute for Applied Systems) was founded. It brought together scientists from both East and West (Cold War) to research on the critical issues of global environmental, technological and social change.

At the European Commission, the Energy Systems Analysis and Modelling programme, closely linked to its task of energy and energy R&D policy formulation, was launched in 1975.

> As a consequence of the difficulties in the world's oil supply the whole energy system has become tight and unstable. All oil dependent countries have therefore to elaborate their future energy policies with much more care and with much less manoeuvre than in the past. In this context, the assessment of the possible impact of energy policies and of related policies has become a high priority exercise. Energy Systems Analysis should play a central role in the assessment process[14].

In 1995, the *Global Scenario Group* was established at the initiative of the Stockholm Environment Institute to examine the prospects for world development in the 21st century[15]. The *Global Scenario Group* recognises the role of deliberate actions and choices (influenced by cultural preferences, social visions and psycho-social factors in shaping the future). This Group associates the unsustainability of the current global trajectory with three critical trends: environmental degradation and resource depletion; increasing income disparity; poverty and marginalisation[16].

The European Commission, an in particular, the *Cellule de Prospective* created by Jacques Delors, was used to elaborate scenarios. Between 1997 and 1999, the Commission Forward Studies Unit produced a set of coherent and thought-provoking images of the future of Europe called "Scenarios Europe 2010"[17]. It presented five qualitative scenarios:

---

14 A. STRUB, *Energy systems analysis*, International Conference of Dublin, 9-11/10/1979, D. Reidel-Kluwer, ECSC-EEC-EAEC, Brussels and Luxemburg, 1980.

15 The Global Scenario Group published the book of P. RASKIN, T. BANURI, G. GALLOPIN, P. GUTMAN, A. HAMMOND, R. KATES, R STEWART, *Great transition: The promise and lure of the times ahead*, Stockholm Environment Institute, Boston, 2002.

16 P. RASKIN, G. GALLOPIN, P. GUTMAN, A. HAMMOND, R. SWART, *Bending the curve: Towards global sustainability*, Stockholm Environment Institute and Global Scenario Group, SEI PoleStar Series Report No. 8, Stockholm, 1998.

17 European Commission, Forward Studies Unit, *Scenarios Europe 2010 – Five possible futures for Europe*, Brussels, 1999.

"Triumphant markets", "The Hundred flowers", "Shared responsibilities", "Creative societies" and "Turbulent neighbourhoods".

In 2006, the EU Institute for Security Studies published "The new global puzzle – What world for the EU in 2025"[18]. This study is divided into three parts: the trends (demography, economy, energy, environment and science and technology), the regions (Eurasia and Russia, Middle East and North Africa, Sub-Saharan Africa, USA, China, India and Latin America) and the EU context (internal challenges, turbulent neighbourhood, and shaping change).

Two main situations can prompt public bodies or private companies to initiate a Qualitative and/or Quantitative outlook:

- *Taking long-term decisions and formulating long-term programmes.* In this long-term perspective, a Qualitative and/or Quantitative outlook is useful for example for setting targets like percentage of greenhouse gas emission reduction or share of renewable energy sources, for planning science and technology funding or for estimating major public spending with long-term implications[19];
- *Coping with major challenges.* Examples of major challenges are the transition in the economic or political system, the demographic change, the competitiveness of a certain territory, the changes to the socio-economic framework (new markets, new legislation etc.), or – more directly linked with energy – with the changes in the natural environment such as climate change.

At the European Union level, for the short and medium-term, the European Commission is used to draw multi-annual planning such as the financial perspectives: "Paquet Delors" for 1988-1992, "Paquet Delors II" for 1993-1999, "Agenda 2000" for 2000-2006 and "Financial perspectives" for 2007-2013. The Commission also uses multi-annual planning for example for its "R&D Framework Programmes" (See Chapter II), for Regional policies (structural funds) or for the Trans-European Networks in the field of energy and transport.

---

[18] N. GNESSOTO, G. GREVI, *The new global puzzle – What world for the EU in 2025?*, ISS, Paris, 2006.

[19] B.R. MARTIN, J. IRVINE, *Research foresight: Priority-setting in science*, Pinter, London, 1989.

Figure III.1: Methods and tools for looking at the futures and for evaluating policies

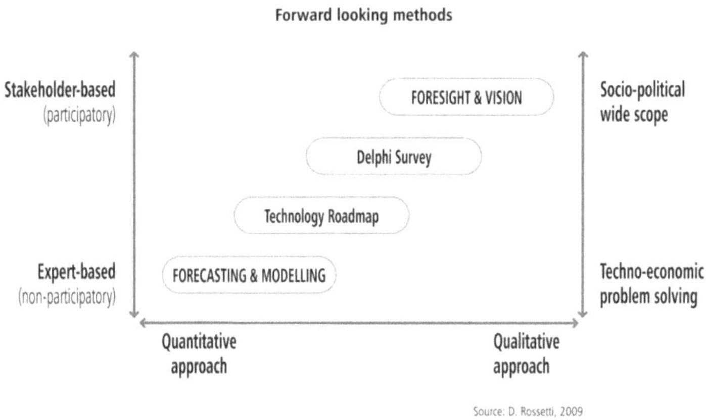

Continuity, exploratory and breakthrough foresight in energy

In the 1970s, Jacques Durand distinguished "Continuity Foresight" (*Prospective de continuité*, i.e. without major breakthroughs from the present trends), from "Exploratory Foresight" (*Prospective exploratoire*, i.e. with adaptation mechanisms to some tensions or clashes), and "Breakthrough Foresight" (*Prospective de ruptures*, i.e. focusing on the uncertainties and on the alternative scenarios rather than on the replication of existing tendencies. Breakthrough Foresight includes the "Normative Foresight" where a sustainable energy and environmental system is defined)[20]. Normative scenarios start by describing a future state of affairs and then asking how this future came about or how one can achieve it. Explorative scenarios start with the present and explore future states by asking "what if" questions[21].

Examples of "Continuity Foresight" can be found in the exercises led by the International Energy Agency (WEO), the European Commission (WETO) or the US Department of Energy (IEO)[22]. With the same horizon time of 20-30 years, these exercises present very similar conclu-

20 J. DURAND, *Prospective, discontinuité et instabilité*, *op. cit.*

21 UNIDO, *Technology foresight manual*, Vienna, 2005.

22 International Energy Agency, *World Energy Outlook (WEO)*, *op. cit.*; European Commission, *World energy, technology and climate policy outlook (WETO-2030)*, *op. cit.*; US Department of Energy (EIA), *International Energy Outlook (IEO)*, *op. cit.*

sions with a world energy demand almost doubling and being largely covered (almost 80%) by fossil fuels.

The energy security of supply is a key issue for the European Union. Today, the EU (27 Member States) is importing 50% of its energy needs (more than 70% for oil, almost 50 for natural gas and 30% for solids). In comparison, the USA import 30% of their energy needs, India 18%, Brazil 15% and China 5%. Among the major economic actors, Japan imports (83%) more than the EU. On the contrary, Russia exports the equivalent of 80% of its total primary energy supply.

The figures mentioned in the 2000 Commission Green Paper and more recently in the Communication "An energy policy for Europe", clearly show that the EU-27 will depend for almost 70% of its energy needs from third countries by 2030 (see Figure III.2). These figures are coming from the PRIMES model simulations[23].

The import dependency is and will be very different among Member States. E.g. countries like France, Lithuania or Bulgaria depend today for about half of their supply from third countries while Germany or Hungary import 60% of their energy needs, and Italy and Ireland for around 85%. Countries like Romania or The Netherlands import 30% of their energy needs and UK 5%.

Figure III.2: European security of supply – EU-27 development of import dependence up to 2030 (Baseline)

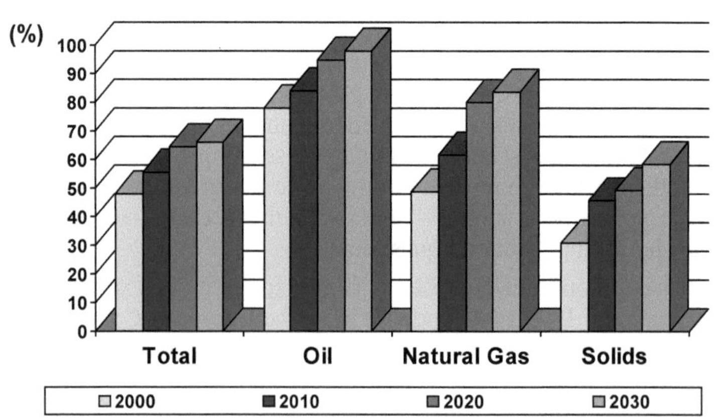

[23] European Commission, *An energy policy for Europe*, COM(2007)1 and *EU energy policy data*, SEC(2007)12.

"Explorative Foresight" can be found in the exercises led by the European Commission or by the Commissariat Général du Plan in France[24]. In 1996, the European Commission presented modelling results for 2020 in four different socio-political scenarios: "Conventional Wisdom", "Battlefield", "Hypermarket" and "Forum". In 1998, the Commissariat Général du Plan published modelling results up to 2020 with some constraints ("*Satisfaire des besoins croissants sous la contrainte des ressources et en respectant l'environnement*") and with alternative scenarios ("*Société de marché*", "*État industriel*" and "*État protecteur de l'environnement*").

Two other reports, one from the International Energy Agency and one from the European Commission can be considered as "Exploratory". They deal with 2050, and are based on modelling (MARKAL and TIMES for the IEA and POLES for the EC)[25]. Interestingly, using different assumptions and different models, the IEA and the EC arrived exactly at the same total energy supply by 2050 (21 Gtoe) in the Reference scenario with more than 70% of the world energy demand covered by fossil fuels. Two other exploratory exercises made for the IEA and the EC within a sustainable development perspective up to 2050 arrived at very similar conclusions using different models (MESSAGE for the IEA and POLES for the EC): the total primary energy supply was reduced by more than 10% and about 50% of the demand was covered by fossil fuels[26].

"Breakthrough Foresight" deals with the long and very long-term exercises (2050-2100) and allows changes of trajectories with major modifications of technologies and behaviours. The Shell scenarios merged qualitative scenarios and simplified modelling to integrate sociological dynamics with institutional and technological breakthroughs. The European VLEEM project combined "Back-casting" with a new energy and environment modelling to take into consideration very long-term socio-cultural behaviours and technological progress[27].

---

24 Commission européenne, *Europe de l'Énergie en 2020 – Une approche par scénario*, OPOCE, Luxembourg, 1996; Commissariat Général du Plan, *Énergie 2010-2020*, La Documentation française, Paris, 1998.

25 International Energy Agency, *Energy Technology Perspectives Scenarios and Strategies to 2050*, OECD, Paris, 2006; European Commission, *World energy technology outlook – 2050 (WETO-H2)*, EC, Luxembourg, 2006.

26 International Energy Agency, *Energy to 2050 – Scenarios for a sustainable future*, OECD, Paris, 2003; European Commission, *World energy technology outlook – 2050 (WETO-H2)*, *op. cit.* – Carbon Constraint Case.

27 *Back-casting* consists in discussing the future from the opposite direction than *Forecasting* that looks at the future on the basis of current trends. In Back-casting, several alternative scenarios are examined – usually representing different normative or desirable states. If one wants to arrive at Scenario A, what trends would need to

The Shell scenarios and the Intergovernmental Panel for Climate Change (IPCC) scenarios are in the mid-way between the "Exploratory" and the "Breakthroughs" Foresight.

The "alternative stories about the future" in the Shell scenarios are based on a qualitative modelling of the technologies and behaviours. The two 2001 Shell scenarios up to 2050 are called "Dynamics as usual" (world driven by efficiency, economic integration and declining power of nation states) and "The spirit of the coming age" that describes "a world in which superior ways of meeting energy needs are developed to meet consumer preferences"[28]. This scenario highlighted the power of cultural values and cohesion. The 2005 Shell scenarios up to 2025 want to take into account the 11 September 2001 events (feeling of insecurity). Three set of forces are provided in the Shell scenarios: market incentives, the force of community, coercion and regulation. The scenarios explore how these forces drive towards efficiency, social cohesion and security[29].

The IPCC released the first global greenhouse gases emission scenarios, the "IS92" scenarios in 1992. These scenarios were essentially modelling oriented. Since then, the IPCC through its Working Group III, developed broader emissions scenarios – more interdisciplinary (and with "open process" – aimed at evaluating climatic and environmental consequences of future emissions of greenhouse gases and for assessing alternative mitigation and adaptation strategies. In 2000, these scenarios – called "SRES" (Special Report on Emission Scenarios) in the third IPCC assessment report – include emission baselines and information on economic restructuring throughout the world. They examine different rates and trends in technological change, they expand the range of different economic development pathways and they include some social developments like the narrowing of the income gap between developed and developing countries. The temporal horizon extends to 2100. In addition to the scenarios driving forces (demographic trends, socio-economic development and direction of technological change) and their relationships, four main storylines (A1, A2, B1, B2) are described and quantified by using several models and approaches[30].

---

change to get him there. Usually, quantitative analysis provides a feasibility or plausibility assessment. For example, if Scenario A was future in which CO_2 emissions were stabilised at Xppm and Scenario B was CO_2 stabilisation at Yppm – quantitative analysis could help assess which one was more plausible under any given set of assumptions.

28 Shell, *Energy needs, choice and possibilities: scenarios to 2050, op. cit.*

29 Shell, *Global scenarios to 2025 – The future business environment: trends, trade-offs and choices*, London, 2005.

30 IPCC, *Special report on emissions scenarios*, Working Group III, Cambridge University Press, 2000.

Figure III.3: Energy Outlooks classification

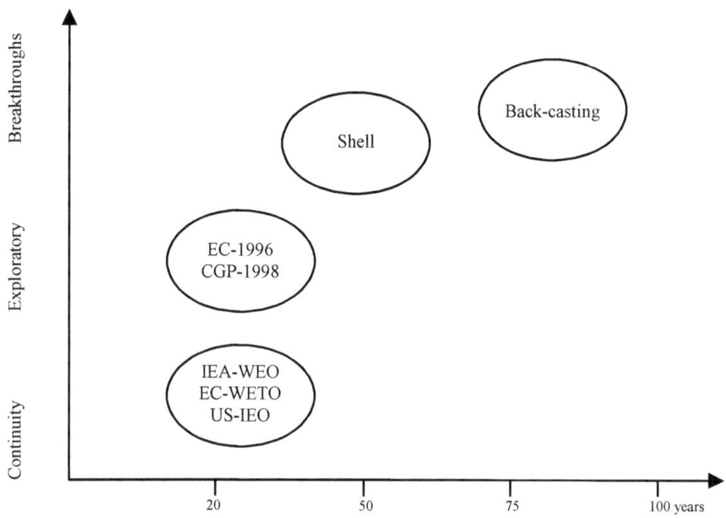

The major issues that the European Union has to face in the energy field are security of supply, competitiveness and fight against climate change[31]. From the beginning of the 1980s, energy in Europe has been a traditional subject to Qualitative and Quantitative outlook (see Chapters II, IV, V, VI and VII). It deals both with long-term decisions and with major challenges. For example, when energy efficiency in the residential sector is concerned, one has to take in mind that to fully renovate the building stock requires up to 100 years; when the electricity sector or the transport infrastructure is concerned 50 years are needed; the industrial equipment has a duration of 10-20 years. Between 3 and 10 years are needed for the rotation of end-use equipment like electrical appliances or cars.

Especially since the end of the 1980s, the energy sector is particularly concerned by the climate change debate. These last years, Kyoto targets and post-Kyoto targets are discussed, directly or indirectly by Heads of State and Governments, in almost all the European Summits.

For the European energy field, two different methods able to explore the futures, to assess the various impacts and to find the ways of achieving a certain goal are investigated:

31 European Commission, *Green Paper on a European strategy for sustainable, competitive and secure energy*, COM(2006)105.

- *Qualitative-oriented outlooks* that encompass the various "Foresight" exercises including Delphi surveys, Critical/Key Technology Studies, Participatory approaches – like the European Awareness Scenario Workshop – and Technology Roadmaps;
- *Quantitative-oriented outlooks* based on "Modelling" essentially the "Top-down" and "Bottom-up" models (See Chapter IV).

Qualitative-oriented outlook

Gaston Berger was one of the first to speak about "Prospective" or "Foresight" that has not to be confused with "Prévision" or "Forecast". In fact, Foresight does not aim to predict the future – to unveil it as if it were predetermined – but to help to build it.

From various authors, one can say that Foresight is[32]:

- *Action-oriented*

Foresight is not only about analysing or contemplating future developments but supporting actors to actively shape the future (studies about the future, for example for advancing knowledge in a specific field are generally not considered as Foresight). Already at the end of the 1950s, Gaston Berger wrote:

> En réalité, hier et demain sont hétérogènes. C'est un regard qu'on jette sur le passé, puisque, de ce côté-là, il n'y a plus rien à faire. C'est un projet qu'on forme pour l'avenir, car là des possibilités sont ouvertes. Passer de la rétrospection à la prospection n'est pas simplement diriger ailleurs l'attention: c'est se préparer à l'action. On peut être prospectif en faisant de l'Histoire[33].

- *Open to alternative futures*

In Foresight exercise the future is not pre-determined and can evolve in different directions which can be shaped by the actions of various players and the decisions taken today. It implies some freedom to choose among the alternative, feasible futures, and therefore increase the chance of arriving at the "selected future".

- *Participatory*

---

32 G. BERGER, *L'attitude prospective*, *op. cit.*; A.C. DECOUFLE, *Traité élémentaire de prévision et de prospective*, PUF, Paris, 1978; Th. GAUDIN, *La prospective*, Que sais-je, PUF, Paris, 2005; H. de JOUVENEL, *Invitation à la prospective*, Perspectives, Futuribles, 2004; H. COURTNEY, *20/20 Foresight: Crafting strategy in an uncertain world*, Harvard Business School Press, Massachusetts, 2002; R.A. SLAUGHTER, *The foresight principle: Cultural recovery in the 21st Century*, Adamantine Press, London, 1995; A. WEBSTER, *Foresight as a tool for the management of knowledge flows*, Report for EC STRATA workshop, Brussels, 22-23/4/2002; J.P. GAVIGAN, F. SCAPOLO, *Matching methods to the mission: A comparison of national foresight exercises*, Foresight, Vol. 1, No. 6, 1999.

33 G. BERGER, *L'attitude prospective*, *op. cit.*

Foresight is not done by a small group of experts or academics. It involves several stakeholders and the results of the Foresight exercise are disseminated to a large audience. The participatory approach is connected to democratisation and legitimacy in political processes.

- *Multidisciplinary*

The philosophy of Foresight is that real problems cannot be understood if reduced to one dimension and sliced up by disciplines.

In a word, Foresight is a non-deterministic, participatory and multidisciplinary approach that can be envisaged as a triangle combining "Thinking the Future", "Debating the Future" and "Shaping the Future". Foresight has been defined by a European high-level group of experts as "A systematic, participatory, future intelligence gathering and medium-to-long-term vision-building process aimed at present-day decision and mobilising joint action"[34].

Figure III.4: Foresight triangle

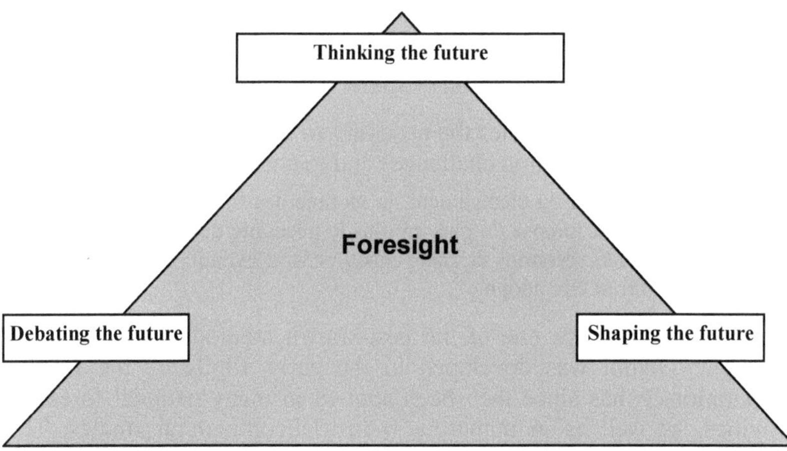

As noted by Paraskevas Caracostas, especially since the 1980s, Foresight tends to "build the future together". It is concomitant to the economic disengagement of the State (cf. Margaret Thatcher). As the State is less present in the planning, it offers to the private and public actors a frame to collectively think on the possible futures ("Futuribles" in French). The objective is to reduce uncertainties and to help actors to define their own strategy. *De facto*, Foresight is not anymore a State planning tool – like in France and the USA after 1945 – but a service for

[34] European Commission, *Thinking, debating and shaping the future: Foresight for Europe*, Final report of the High level expert group, 24 April 2002.

the actors themselves. The spirit is linked to the new "governance" understood as coordination of heterogeneous actors from the public, private and associative sectors[35].

This way of making Foresight in the context of public policies is particularly clear in the UK where the Prime Minister has a "Strategy Unit" that works together with the foresight units in the different ministerial departments[36]. In Finland, a "Committee for the Future" has been established at the Parliament. In addition, the Finnish Government should communicate its view on scenarios and strategic challenges to Parliament once each electoral cycle. This should provide a basis for political and public debate, and increase scrutiny of the extent to which individual policies are consistent with the Government's wider strategy.

Jean-Marie Chevalier defines the "scenario method" as a collective reflection aiming to describe different pictures of the future.

> L'exercice consiste à organiser une réflexion collective visant à décrire différentes images de ce que pourrait être l'avenir (en l'occurrence énergétique), à un horizon temporel qui peut être de dix, vingt, trente, cinquante ans. Il ne s'agit pas de prévoir mais de décrire différentes images possibles (…). L'exercice met en évidence les incertitudes, les risques auxquels une entreprise ou une nation peuvent s'exposer[37].

Peter Schwartz underlines the necessity to have a large set of scenarios to face the globalisation challenges and to choose the best strategy.

> Avec l'accélération du changement, la montée des interdépendances et des incertitudes, s'est impose de plus en plus la nécessité de prendre en compte une multitude de scenarios et, dans chaque cas, d'examiner quelle stratégie optimale pourrait être adoptée[38].

Delphi surveys are one of the best-known methods for technology Foresight. Delphi was developed in the early 1960s by the RAND Corporation. It has since then been applied in many national foresight activities, as well as in numerous industrial/commercial studies. The core of the Delphi method is a multi-round survey. With each questionnaire, the participating experts receive the results from the previous rounds. Thus, the participants are asked to reconsider their judgments based on the opinions of their peers.

---

35 P. CARACOSTAS, European Commission, DG Research, *Briefing Prospective*, internal note, 16/11/2007.

36 UK House of Commons, *Governing the Future*, second report of session 2006-2007, London, 2007.

37 J.M. CHEVALIER, *Les grandes batailles de l'énergie*, *op. cit.*

38 P. SCHWARTZ, *La planification stratégique par scénarios*, *op. cit.*

The European EURENDEL project funded in the 5th RTD Framework Programme was the major EU-wide Delphi study on social and technological visions in the energy sector. Dealing with the 2030 time horizon 2030 and the enlarged European Union (EU 25), two Delphi rounds were held with more than 3400 invited experts (680 responses in the first round and 470 responses in the second survey round)[39]. The main identified problem fields were: hydrocarbon bottleneck; grid development; storage technologies; renewable fate; environmental objectives; energy market, prices and taxes; future social relations, future of work, demographic trends; transport and mobility spatial movements; technological progress; nuclear power; future demand and efficiency. Twenty technological statements ("Novel production processes", "25% biofuels", "Superconductive materials", etc.) and three societal visions ("Individual choice", "Ecological balance", "Social equity") were the basis of the Delphi survey.

Among the results of Delphi survey the following points could be highlighted:

- Experts answering to the Delphi survey were significantly more "optimistic" than EU reference quantitative studies in terms of potential technological breakthroughs. E.g. the statements "H2 from Renewables" significantly contributing to the energy systems or "5% PV" for EU electricity by 2050 were respectively about 20 times and 10 times more "positive" in the EURENDEL experts responses than in the quantitative study made with PRIMES for the European Commission[40].
- Energy demand technologies ("novel production processes" and "intelligent buildings") receive the highest scores in terms of social impacts: they were always preferred.
- Short-medium term time of occurrence, i.e. before 2025: "Biomass widely used for heating", "15% freight on rail", "30% distributed energy systems", "energy storage in Renewables", "CO_2 capture". The other statements should occur after 2030.
- Close to all (i.e. at least 95%) experts say that the following statements will occur around 2025: "intelligent buildings", "20% fuel cell cars", "H_2 from diverse sources", "Energy storage in RES", "Superconductive materials", "25% RES".

---

39 The EURENDEL participants were IZT (Timon Wenhert), PROSPEKTIKER (Daniela Velte), EC-BREC (Anna Oniszk-Poplawska), Università Bocconi (Augusto Ninni) and RISOE (Birte-Holst Joergensen).

40 European Commission, *Energy and transport trends to 2030*, OPEC, Luxembourg, 2003 and 2006.

Critical Technology studies are usually based on interviews with industrial experts on the technology for which a forecast is being developed. These studies aim to highlight short-term (3-10 years) R&D priorities for policy makers[41]. They apply sets of criteria against which the importance or criticality of particular technologies can be measured. Sometimes, a benchmarking analysis provides comparisons with other technologies considering that Benchmarking refers to the process of measuring and evaluating products, services and technologies in comparison to competitors that are considered the best ones. This process should allow learning from best practices elsewhere. No *Critical Technology studies* have been led in the energy socio-economic research of the EU RTD Framework Programmes.

Participatory approaches build on the idea that knowledge is distributed widely and that no single public body or private organisation can know everything that needs to be known in order to effect the desired changes. More used in the Scandinavian countries or The Netherlands, Participatory approaches encourage the participation of individuals and groups[42]. Future-oriented activities can be considered participatory when:

- They involve participants from different stakeholder groups (e.g. experts and policy-makers, researchers and business men, etc.);
- They disseminate their preliminary results (analyses, tentative conclusions, policy proposals, etc.) among interested "non-participants" (face-to-face at workshops, over the Internet with free access for everyone, or in the form of printed documents, leaflets, newsletters);
- They seek feedback from this wider circle[43].

The European SIREN project, funded in the 5th RTD Framework Programme, aimed to disseminate and implement European and national R&D results in the field of renewable energy through the so-called European Awareness Scenario Workshop (EASW). This EASW methodological approach was developed to raise public awareness, to build consensus on new technological solutions, to activate a participatory process among stakeholders and to promote public acceptance. SIREN focused on four European Municipalities (Leicester in the United

41 Ministère français de l'économie, des finances et de l'industrie, *Technologies clés 2010*, Paris, 2006.

42 A. SALO, R. GUSTAFSSON, I. KAAKKOLAMMI, T. GUSTAFSSON, *Participatory development of future-oriented policy configurations for climate change and mobile gaming*, 2004.

43 An example of Participatory approach is the *European Awareness Scenario Workshop* (EASW).

Kingdom, Lisbon in Portugal, Deventer in The Netherlands and Spoleto in Italy)[44].

The first part of an EASW is oriented towards the articulation of a "common vision". Four scenarios, outlining four divergent visions of what the city might be like are presented, depending on certain policy decisions and available options. During the first session, generally the Mayor gives an introduction about the present state of affairs regarding urban ecology in the city where the workshop takes place.

In a second phase, the participants are split up into four role groups (residents, politicians and civil-servants, technical experts and business sector) that articulate their own future vision of their "sustainable future". This results in four future visions that are then presented and discussed in a plenary session. A list is drawn up of all the elements the four visions have in common. This forms the basis for a commonly shared future vision, the so-called "common ground".

The second part of the EASW is oriented towards generating ideas on how the participants' shared vision of a sustainable city might be realised. Taking place in theme groups – originally focusing on energy, water and wastewater, waste and recycling, living – each participant write down its best ideas towards accomplishing sustainability in terms of the topic allocated to the theme group – one idea per small card (a "snow card"). This should result in 5-10 ideas per person that are then prioritised in a short list of the "best" 5 ideas produced in the theme group.

The plenary session that follows begins with the presentation of the four "top-5" lists. The main purpose of each presentation is to outline the group's ideas and indicate their relevance to the "common ground vision". After the four presentations, the overall "best 5" ideas are identified using a voting procedure. The closing discussion then focuses on the "winning" ideas, their merit, their feasibility, and how they might be realised. The solutions are framed in two dimensions which, when combined into a cross, give four possible answers to the central questions of WHO and HOW.

Who is responsible for solving local problems? Is the public authority or the private sector? How should technology be used in solving the problems? Should we look to people or technology for a solution to the problems of sustainability? Each solution is analysed by building a diagram/matrix based on two different axes/ dimensions: technology content (low-tech or high-tech solution) and type of solution (public or

44 The SIREN participants were INNOVA, Agencia per l'Energia e l'Ambiente (Perugia), Lisbon Municipal Energy Agency, NOVEM, Leicester Energy Agency, University of Lisbon, Dialogic and Ecoazioni.

private) which will allow defining how is the community orientation towards technology or social solutions.

Finally, an Action Plan is defined on how and who is making what in order to achieve the common shared vision. Sometimes, the Mayor (or other stakeholders) signs this Action Plan as a commitment

Sociologically oriented (importance of the role of each single stakeholder, influence of the group, etc.), the EASW have three positive elements:

- Participatory by essence, EASW raises the awareness of the "multiplers" in the local community on a specific field like sustainable development and renewable energies in the case of SIREN;
- EASW allows to highlight the existence and the potential of new technologies (like renewables in SIREN), i.e. of emerging options like than only relatively mature technologies. It is interesting to notice than out of one hundred Coordinators of research projects, only a few were interested to try to exploit directly their result.
- This participatory approach creates synergies between public and private actors, and increase the exchange of knowledge between technological, legal and socio-economic actors ("this kind of Photovoltaïc panel has already been demonstrated, this decree provide such subsidy, this technology allows to save X€ after Y years").

On the question marks of this methodology, on can raise the issues of the:

- Relevant political level (European or, more probably, regional or local) for funding such initiatives also considering the subsidiarity principle;
- The justification of the cost of the initiative in view of the uncertain achieved results as only the raised awareness of local stakeholders is guaranteed;
- The share of "research" and the share of "dissemination" in such kinds of initiative.

Figure III.5: European Awareness Scenario Workshop (Participatory methodology)

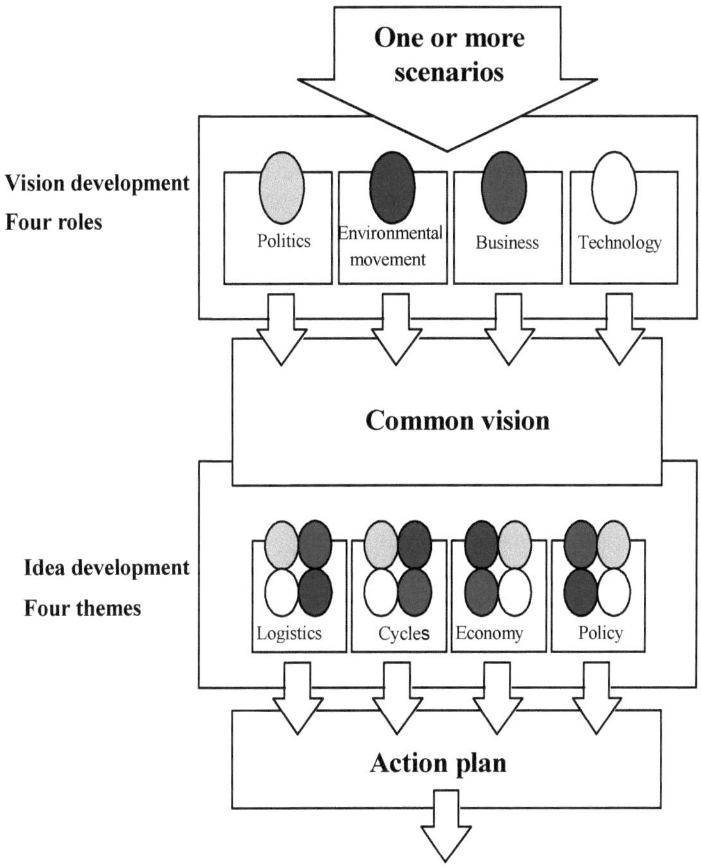

A *Technology Roadmap* is "an extended look at the future of a chosen field of inquiry composed from the collective knowledge and imagination of the brightest drivers of change in that field"[45]. A Roadmap includes graphical representations in which "nodes" (past, present or future states of the art in S&T development) are connected by "links" (causal or temporal relations) showing the nature, rate and direction of potential S&T developments from or towards those nodes. These graphical representations help in informing decisions about possible future options. As such, a roadmap is also a planning methodology. It is

45 R. GALVIN, *Science roadmaps*, Science, Vol. 280, No. 5365, 1998.

a "traveller's tool that provides essential understanding, proximity, direction, and some degree of certainty in travel planning"[46]. Technology Roadmaps are often developed and used by large high-tech firms, like in the semiconductor industry[47].

> Motorola has prolifically used sophisticated engineering roadmaps to great advantage over several decades (…). Roadmaps allow our industry leaders to communicate convincingly with those in government and business regarding their support of our goals[48].

One European energy socio-economic research project – ERMINE for the electricity sector and funded by the sixth EU RTD Framework Programme – dealt with a Technology Roadmap[49]. Nevertheless, the majority of the so-called "European Technology Platforms" led by industry have as a first task to establish a Roadmap, sometimes called a "Technological common vision"[50]. In a second phase, the Technology Platforms should provide a strategic research agenda setting the medium and long-term objectives for the technology. Finally, the Technology Platforms have to implement the research agenda by mobilising the significant human and financial resources. In the energy field, since 2000, six Technology platforms have been established on Photovoltaics, Biofuels, Wind, Solar Thermal, Smart Grids, Zero-emission fossil fuel plants, and Hydrogen and fuel cells[51].

---

46 R. KOSTOFF, R. SCHALLER, *Science and technology roadmaps*, IEEE Transactions on Engineering Management, 2001.

47 W. SPENCER, L. WILSON, R. DOERING, *The semiconductor technology roadmap*, Future Fab International 18, USA, 12 January 2005.

48 R. GALVIN, *Science roadmaps*, Science, *op. cit.*

49 The ERMINE consortium was coordinated by CESI (Michele De Nigris) with the following partners: KEMA (Nanno Bolt), ENEL (Pietro Barbuccci), EURELECTRIC (Mihai Paun), EBL (Aril Fogh Thommesen) and IPE (Madajewski Krzysztof).

50 European Commission, *European Technology Platforms – Knowledge for Growth*, EC, Brussels, 2005.

51 European Commission, *Hydrogen energy and fuel cells – A vision of our future*, EC, Brussels, 2003.

Figure III.6: European Hydrogen and Fuel Cells Roadmap

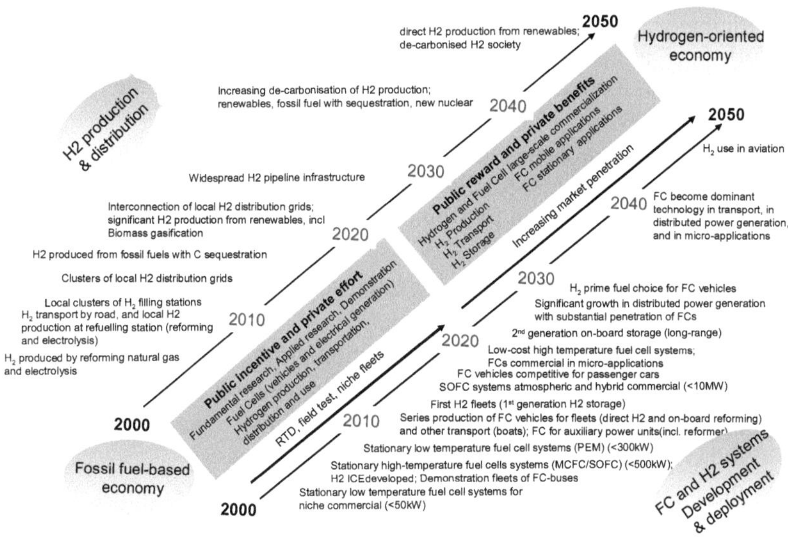

Strengths and weaknesses of qualitative-oriented outlooks

The following main strengths can be derived from the qualitative exercises:

- Foresight allows the strengthening of communities of experts concerned by technology futures and technology watch activities; this has positive impact in terms of multidisciplinarity and for building networks among scientists, industry, public authorities and non-governmental organisations.
- Foresight generally involves both experts and key stakeholders in an open and interactive process.
- Prior information and treatment of the information received by the participants facilitates informed discussion.
- Delphi could be considered an anonymous group discussion. While introducing some of the advantages of a discussion into the survey, the anonymity of the process ensures that the opinions of influential individuals do not dominate the findings. Instead the most likely or convincing developments are identified based on rational arguments.
- The underlying hypothesis of the Delphi method is that experts in a field have pictures of likely future developments in their head (expert knowledge). By asking a larger number of experts, individual mis-

perceptions are taken out. In addition, by applying feedback loops the soundest predictions become visible.

- Participatory activities like the "European Awareness Scenario Workshop" are particularly efficient in local context where it is relatively easy to associate to the problems the actors having the responsibility to solve it.
- Participatory approaches have been initially experimented in the environmental field, especially for the solution of typical urban problems. They can be useful to promote more sustainable models shared by the community.

The following Weaknesses have been identified from the Qualitative-oriented outlooks:

- Difficulties in finding an optimal selection process of:
  - the topics to be addressed, for example why "Mobility and Communication" and "Health and Quality of Life" were selected in the German Futur exercise[52]?
  - the experts (futurologists/futurists) that choose the topics under discussion. The experts selected are often from the "community" of the people making the Foresight exercises.
  - the participants to the workshops: weight of the different stakeholder groups and distinction between the "inner circle" (active participants who took part in the workshops, in the focused groups, etc.), and the "outer circle" (people invited mainly to vote or to contribute via Email or Internet).
- Risks of bias coming from the "Expert presentations" to the workshops or to the focused groups.
- Foresight embraces sometimes too large subjects and is not sufficiently targeted on specific issues. *De facto*, many results do not provide sufficient details for specific policy purposes (figures in a Directive, economic analysis, etc.). as it will be demonstrated in Chapter VI.
- The outcomes of the workshops or the focused groups are drafted by the specialists.
- Public relations and communication could bias the selection of the topics (if the public interest is higher for nuclear and wind, why selecting coal and natural gas?).

---

52 Bundesministerium für Bildung und Forschung, *Futur: Der deutsche Forschungsdialog*, BMBF, Bonn, 2003.

- Delphi set-up tends to develop mainstream or business as usual answers; the creativity of outsider opinions risks to be lost and interruptions of the system are not accounted for[53].
- The choice of experts is too often technology oriented. Consequently social developments and needs are not appropriately considered and the technological potential is sometimes overestimated by professional optimism.

53 A prominent example in the energy field is the oil crisis of the seventies which rendered quite few of the predictions made in the sixties worthless.

CHAPTER IV

Economic and Technology Modelling

Summary

After having analysed, categorised and compared the Qualitative and the Quantitative outlooks (Chapter III), this Chapter firstly highlights the characteristics of Models and the representation of the energy system from an economic (Top-down) or a technological (Bottom-up) perspective. The objective is to underline some of the major strengths and weaknesses of economic and energy models.

Models attempt to reproduce the real world, are normative and aim to elicit the best-available policy or technological options. Top-down models are strong in terms of economic explicitness (micro-economic realism and macro-economic completeness) and Bottom-up models are technology-rich. Top-down models are particularly useful for studying the interactions between the energy system and the rest of the economy whether Bottom-up models describe energy technologies in details from both supply and demand-sides.

All along this part, a specific focus is on the European Union energy economic research (see Chapter II). In particular, the main features are given on the main models developed at the EU-level from EFOM and MEDEE at the beginning of the 1980s to POLES and PRIMES today in the field of energy, from E3ME and GEM-E3 to NEMESIS in the economic field. Comparisons are also provided with some models developed in the USA (NEMS model) or in some countries (MARKAL).

Secondly, the main critical issues in modelling are addressed. They concern the database access (difference between Top-down and Bottom-up models), the basic assumptions ("techno-socio-economic paradigm") and the exogenous parameters like the economic and demographic growth, the constraints on energy resources or the technological development.

Another critical issue of modelling, especially concerning the communication of results to policy-makers and the public, regard the comparison of the projections with a Reference year or with a Reference case. Sociological aspects ("models influence policy-makers or policy-makers influence the models") and technical (for example for hydrogen or for CO_2 capture and storage) and geographical representations (for example of emerging and developing countries) are the last two issues raised in this part of this chapter.

Thirdly, the long-term (up to 100 years) and the "Back-casting" approach are analysed. The questions of uncertainties (economic and technological development), of "tyranny of the present", of structural changes – including socially and institutionally – and of intergenerational welfare are addressed. Some results of the VLEEM European project are presented.

Finally, the issues of technological progress and policies for mitigation of climate change are addressed. The learning by doing concept, the endogenous technical change in Bottom-up and Top-down models as well as several EU energy economic projects results and literature review are discussed. The main issues concern the learning/ experience curves, the induced technical progress, and the cost of reducing emissions of greenhouse gases.

Examples of EU energy-related policies (European Council targets, and results of modelling activities illustrate this Chapter. Critics and ways of methodological improvements in modelling, essentially coming from a study made by Dominique Finon, are also part of this Chapter.

Among the main conclusions of this Chapter are the economic validity of the Top-down model exercises and the technological robustness of the Bottom-up model analyses. But some major issues have still to be improved in modelling like the interaction between the economic and the energy systems or the treatment of technological progress in particular in a long-time perspective.

Top-down and Bottom-up Models

If Forecasting is the process of predicting the future based on current trend analysis, one can say that Modelling is

Attempting to reproduce the real world

Modelling is the attempt to reproduce real-world defining and quantifying relationships using mathematics, statistics and algorithms. Like Systems analysis and Operations research, it deals with the analysis of complex, large-scale systems and the interactions within those systems within a computer model. It aims to form a detailed overall picture.

Normative

Modelling looks at future (technological) developments based on the extrapolation of perceptible trends. It may involve the use of time-series data from the past to extrapolate the likely path of future trends or the possible occurrence of future events.

Eliciting the best-available policy or technological option

Modelling should allow eliciting the best possible solution to a problem.

The energy security of supply issues in the 1970s and the climate change issue in the 1980s have spurred the creation of different Economics, Energy and Environment (E3) models.

Two main approaches exist for modelling the so-called E3 interaction: the "*Top-down*" and the "*Bottom-up*" models. They differ mainly with respect to the description of the general economy and to the emphasis placed on the technological details of the energy system[1].

The "*Top-down*" macro-econometric models describe the energy system (as another sector) in an aggregate way, by means of neoclassical production functions that capture substitution possibilities by means of substitution elasticities (the assumptions of factor and price flexibility was first theorised by Jean-Baptiste Say in the 19th century)[2]. The elasticities of substitution define how easily one aggregate input can be substituted for another as their relative prices change. The other key parameter is the autonomous energy efficiency index – estimated from historical data – that reflects an assumed degree of decoupling between GDP and energy consumption[3].

Top-down models capture feedbacks effects of energy policies on non-energy markets such as price changes for factors or intermediate goods. They provide a consistent framework for studying price-dependent interactions between the energy system and the rest of the economy (cf. spillover effects, GDP, employment or welfare impacts). Macro-economic models simulate energy demand trends and the impacts of energy sector changes on the economy, based on historical

---

1 J.P. BRUCE, H. LEE, E.F. HAITES, *Climate change 1995: Economic and social dimensions of climate change*; Contribution of Working Group III to the Second Assessment Report of the IPCC, Cambridge University Press, UK, 1995. P. CRIQUI, L. VIGUIER, *The costs of CO_2 reduction and the impact of flexibility mechanisms in meeting the Kyoto targets*, Int. J. Global Energy Issues, 2000. M. GRUBB, J. EDMONDS, *The cost of limiting fossil fuel CO_2 emissions – a survey and analysis*, *op. cit.*

2 J.-B. SAY, *Traité d'économie politique*, Paris, 1803.

3 I.S. WING, R.S. ECKAUS, *The implications of the historical decline in US energy intensity for long-run CO_2 emission projections*, Energy Policy, Special edition, Technology in climate policy and climate models, ed. Ph. THALMANN and L. VIGUIER, Vol. 35, No. 11, November 2007.

relationship between energy-use, energy prices and various economic, demographic and other parameters.

> Les modèles d'équilibre général calculables décrivent explicitement les comportements micro-économiques des agents en fonction de leurs préférences, de fonctions de production qui représentent les techniques, et équilibrant offres et demandes par le système des prix[4].

The Top-down models have been dominated by Computable General Equilibrium (CGE) models but the Keynesian macro-econometric models – that model mostly the demand – are often preferred for the short-term time horizon. In the USA, Dale Jorgenson's modelling approach and numerous similar models have dominated policy analysis.

Speaking about the French experience using macro-econometric models, Paul Zagamé highlights that the different economic models have a similar architecture but different mechanisms and that could help to frame the real economy ("C'est peut-être au prix de cette diversité que l'on peut espérer *encadrer* la réalité économique"). He also insisted on the various uses of the model:

> Ces utilisations sont variées, prévisions, simulations variantielles analytiques d'une mesure de politique économique, élaboration de scénarios optimisés, réinterprétation quantitative précise de l'histoire économique récente[5].

The major examples of Top-down models developed at the EU-level are the general equilibrium model GEM-E3 or the econometric models NEMESIS and E3ME. One difficulty for the econometric models – estimated on past behaviours – is to envisage radical different behaviours from the economic agents like a social paradigm shift. Concerning climate change policies, Top-down models – most often in a computable general equilibrium framework – allow calculating the greenhouse reduction costs in terms of GDP losses for different levels of reduction.

The "*Bottom-up*" engineering models have been among the first ever-constructed energy system models for the European Union. For example, the EFOM model was built at the beginning of the 1980s. EFOM used linear programming to optimize the primary energy supply needed to meet a given level of final demand. At that time, the MEDEE techno-economic model was also used to simulate energy demand at the final consumption level – industry, transport, residential and tertiary

---

4 D. FINON, *Prospective énergétique et modélisation – Identification de pistes de progression méthodologique*, *op. cit.*

5 P. ZAGAME, *L'expérience française de modélisation macro-économétrique: bilan et perspectives*, *op. cit.*

sectors (See Chapter VI)[6]. Similar technology-rich models are MARKAL developed in many countries or NEMS for the USA.

Bottom-up models focus on the potential for advanced energy technologies to replace existing technologies with competitive increases in efficiency and profitability. Such models have formulated a single, global mathematical programming problem that covers the energy system. Bottom-up models seek to optimise energy supply and demand based on energy policy scenarios that mobilise the unrealised energy efficiency potential of technologies[7]. Concerning climate change policies, Bottom-up models allow building greenhouse gas emission reduction curves for different marginal costs.

Since the beginning of the energy system analysis in the 1970s, Bottom-up and Top-down approaches have proceeded along separate paths. For example, Michael Grubb, Jae Edmonds, Patrick Ten Brink and Michael Morrison highlighted the energy efficiency gap in the 1980s: Bottom-up models suggested the existence of significant no-regrets possibilities for increasing energy efficiency in the economy; Top-down models assumed that competitive markets automatically allocated inputs and final goods efficiently, i.e. denying the existence of an energy efficient gap[8].

In the 1990s, these two modelling "tribes" – in the terminology used by Jean-Charles Hourcade – began to engage in a productive dialogue[9]. They were stimulated by the need to cope with the liberalisation of the energy market in Europe and to manage climate change (e.g. to evaluate the costs of CO_2 mitigation policies or to assess the potential of a certain technology).

---

6 B. CHATEAU, B. LAPILLONNE, *Energy demand: facts and trends – a comparative analysis of industrialized countries*, Springer-Verlag, Vienna, 1982. Studying the dynamic components in the modelling of energy consumption, Jacques Girod concluded that "The process of permanently adjusting the real demand to the long-term desired demand leads us to connect the demand at time *t* to that observed at *t*-1, implying the hypotheses of a 'memory' in the modes of consumption because of users' habits and of the rigidity of their equipment stock". J. GIROD, *Dynamic demand analysis and the process of adjustment*, in J.H. KEPPLER, R. BOURBONNAIS, J. GIROD (eds.), *The econometrics of energy systems*, *op. cit.*

7 One of the first global application of a Bottom-up model was published at the end of the eighties: J. GOLDEMBERG, T. JOHANSSON, A. REDDY, R. WILLIAMS, *Energy for a sustainable world*, Wiley, New York, 1988.

8 M. GRUBB, J.A. EDMONDS, P. TEN BRINK, M. MORRISON, *The cost of limiting fossil fuel CO_2 emissions: A survey and analysis*, *op. cit.*

9 J.C. HOURCADE, *Les modèles dans les débats de politiques climatiques: entre le capitole et la roche tarpéienne?*, in A. DAHAN DALMEDICO (ed.), *Les modèles du futur – Changement climatique et scénarios économiques: enjeux scientifiques et politiques*, Recherches, La Découverte, Paris, 2007.

As Bottom-up models were lacking explicit representation of market-based policy instruments and of realism in formulating the "individual" behaviour of demanders or suppliers, market-oriented modelling approaches have been developed. These models often called "new generation models" are characterised as partial equilibrium models because they cover only the energy system and not the rest of the economy.

POLES for the world and PRIMES for Europe are such kind of partial energy Bottom-up engineering-based models. They incorporate many energy technologies in detail. They combine engineering-orientation with economic market-driven representations. For this latter point, they have some common features with the "Top-down" models as prices play a key role in the adjustment of most variables in the model.

At the EU-level, models like GEM-E3, POLES or PRIMES allow addressing medium to long term issues (about 30 years time horizon). They provide endogenous mechanisms that illustrate long-term substitutions among production factors, changes in consumer patterns, changes in technology, changes in trade patterns and investment that would be driven, for examples, by carbon constraint measures (climate change targets) or by long-run expectation of higher fossil fuel prices. Several EU energy, economic and environment analyses will be carried out with these models as it will be shown in Chapter VI on "Case studies on European energy-related policies[10]. The European Commission evaluated that the cumulative energy investments amounted to € 1,800 Billion between 2005 and 2030 and mostly in the electricity sector (mostly generation and about 35% for distribution and transmission)[11].

---

10 As it will be described in Chapter IV, rather than "models", SAFIRE and MURE that will be used in EU policy-making are database-oriented softwares for assessing impacts of new energy technologies (like renewables) and of energy efficiency policies. They measure the impacts of a higher share of renewables or of a specific energy efficient measure (labelling, standards, etc.) at the local, regional, national or European level.

11 European Commission, *Staff working document on EU energy policy data*, SEC(2007)12.

Figure IV.1: EU-27 cumulative energy investments (2005-2030): € 1,800 Billion

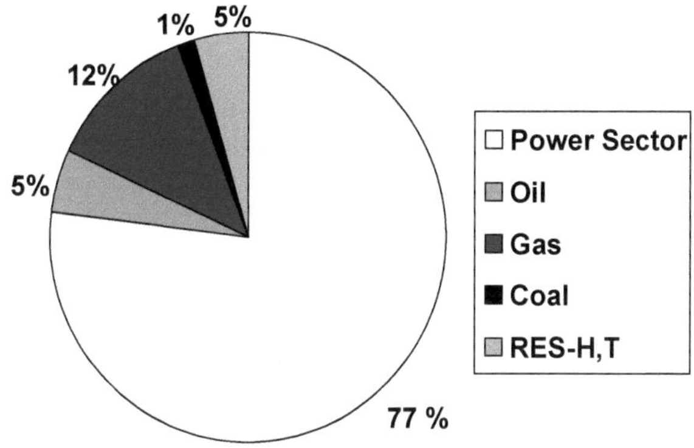

Jean-Charles Hourcade, Mark Jaccard, Chris Bataille and Frédéric Ghersi focused on the macro-economic completeness and micro-economic realism of conventional Top-down models and on the technological explicitness of conventional Bottom-up models to promote better, "ideal" or "*Hybrid models*"[12]. Such kind of models is being developed in two ways: either by incorporating macro-economic feedbacks in the Bottom-up models[13]; either by adding technological explicitness to a Top-down computable general equilibrium model[14].

The first case was done, for example with the MARKAL optimisation model or with a technology-rich and behaviourally realistic model known as CIMS model[15]. In the second case, this is done by representing

[12] J.C. HOURCADE, M. JACCARD, C. BATAILLE, F. GHERSI, *Hybrid modelling: New answers to old challenges*, *op. cit.*

[13] C. BATAILLE, M. JACCARD, J. NYBOER, N. RIVERS, *Towards general equilibrium in a technology-rich model with empirically estimated behavioural parameters*, The Energy Journal, Special issue, IAEE, Cleveland, 2006.

[14] C.C. KOOPMANS, D.W. TE VELDE, *Bridging the energy efficiency gap: Using Bottom-up information in a Top-down energy demand model*, Energy Economics, No. 23, 2001. J. MC FARLAND, J. REILLY, H. HERZOG, *Representing energy technologies in Top-down economic models using Bottom-up information*, Energy Economics, No. 26, 2004. A. SCHAFER, H. JACOBY, *Technology detail in a multi-sector CGE model: Transport under climate policy*, Energy Economics, No. 27, 2005.

[15] A. MANNE, C.O. WENE, *MARKAL-MACRO: A linked model for energy-economy analysis*, Brookhaven National Laboratory, 1992; M. JACCARD, A. BAILIE, J. NYBOER, *CO_2 emission reduction costs in the residential sector: Behavioral parameters in a Bottom-up simulation model*, The Energy Journal, No. 17, 1996;

explicit technologies as Leontief fixed input share production functions and allocating market share amongst these technologies using life-cycle cost competitions or by advanced mathematical techniques[16].

In order to face the Top-down and Bottom-up models drawbacks, D. Finon highlighted several potential methodological progresses[17]:

- Establishing relations between Bottom-up and Top-down models to improve the representation of substitution possibilities among factors or among consumption goods (price changes);
- Definition of *Hybrid models* (see below) with fine tuning representation of the production functions of energy industries;
- Formalisation of endogenous technological progress in the Bottom-up and Top-down models like innovation adoption dynamic effect or like the impact of public R&D;
- Quantified transcription of the effects of the new innovation wave associated to the new Kondratieff cycle in the representations of interindustrial relations in Top-down models to evaluate the cost of policies in the future technical and environmental framework.

C. BATAILLE *et al.*, *Towards general equilibrium in a technology-rich model with empirically estimated behavioural parameters*, *op. cit.*

16 C. BOEHRINGER, *The synthesis of Bottom-up and Top-down in energy policy modelling*, *op. cit.*; R. SANDS, *Dynamics of carbon abatement in the second generation models*, Energy Economics, No. 26, 2002.

17 D. FINON, *Prospective énergétique et modélisation – Identification de pistes de progression méthodologique*, *op. cit.*

Figure IV.2: Energy system in Top-down and Bottom-up models

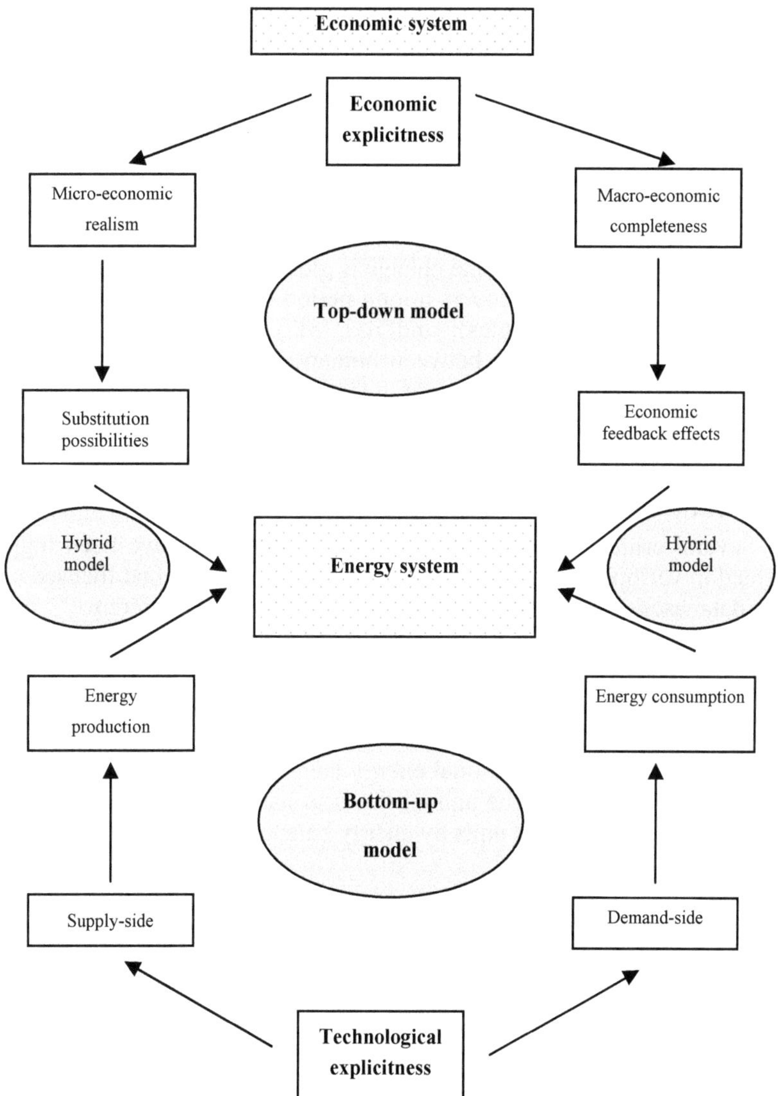

Without entering in the details, one has to mention the existence of "*Integrated (Assessment) models*". These models have been developed in the context of global climate change. Collection of information from different disciplines (essentially physics and economics) – these models should allow evaluating alternative policy options that require a broad look across all the elements of the climate problem (carbon cycle, change of seawater temperatures, energy demand, etc.)[18].

Involving different analytical approaches in different parts of the problem, integrated models combine impact analysis and cost estimates. In developing integrated models, Hadi Dowlatabadi started from the following assumptions: climate change is global (and requires a globally coordinated response); it covers a long period of changes (much longer than normal political cycles); and it is very complex (in particular looking at the interactions between human activities, the atmosphere, and natural ecosystems). Therefore, integrated models were designed to explore and understand these interactions and evaluate ways to avoid, mitigate and/or adapt to global climate change[19].

Critical issues in modelling

Several critical issues exist in modelling and they have been highlighted in various Forum. In this section, the attention will be focused on the databases, assumptions, exogenous parameters, "Reference", discount rate, past results, sociological aspects and technical and geographical representation.

Database

The methodology of national energy balances has been developed at the end of the 1950s[20]. The annual flows in national energy system are reported in homogeneous units by energy balances in a standard format.

---

18 Models (Top-down and Bottom-up) for simulating greenhouse gas abatement policies are still substantially less complex than most of the natural science models simulating the global climate system. As recently suggested by the World Modelling Summit for Climate Prediction, "Climate models will play an important, and perhaps central, role in guiding the trillion dollar decisions that the world will be making to cope with the consequences of changing climate". ECMWF, *World Modelling Summit for Climate Prediction*, Reading (UK), 6-9 May, 2008.

19 DOWLATABI H, *Integrated assessment of climate change: An incomplete overview*, in N. NAKICENOVIC, W.D. NORDHAUS, R. RICHELS, F.L. TOTH, *Integrative assessment of mitigation, impacts, and adaptation to climate change*, IIASA, Laxenburg-Austria, 1994.

20 National energy balances include three sections: primary energy supply, transformations, final energy consumption). Sections on transformation to energy services and energy services provided do not exist. The technological dimension of the current national energy balances could be improved in the transformation section.

At the EU-level, since the 1970s, Eurostat compiled national energy balance extended to end-uses. At this time, some countries developed the energy version of Input/Output matrixes. Recently, additional information relevant for the energy sector such as prices or emissions are reported unrelated to national energy balance.

Giancarlo Tosato highlighted that Top-down modellers can access several databases of econometric data. The international GTAP (Global Trade Analysis Project) computable general equilibrium model has been created to improve the quality of quantitative analysis of global economic issues within an economy-wide framework. The centrepiece of GTAP (covering 96 regions and 57 sectors) is a global database describing bilateral trade patterns, production, consumption, and intermediate use of commodities and services.

On the contrary, Bottom-up modellers have to spend much more time at the onset of their modelling effort because so far no such standard initial database is available and has to be developed on purpose.

Integrating the concept of national energy balance with the concept of energy technology databases would help to achieve more reliable quantitative information on energy systems. In fact, it would allow to cross-check energy flows with the stock of capital in the supply (power plants, refineries? etc.) and end-use sectors (refrigerators, lamps, cars, etc.)[21].

Assumptions

Any constructed system of thought is based on assumptions that cannot be verified for ultimate truthfulness and must be accepted on faith. The paradigms that are built on these assumptions are the foundations of the way we think and act[22]. Closer to energy modelling, Dominique Finon speaks about the need for modellers to disclose the emerging "techno-economic paradigm" that is not anymore largely depending from a new energy form but is more related to its environmental impacts and to the concerns about limited resources and climate change[23].

---

21 G. TOSATO, *Applied systems analyses, technology and research*, Energy Models, Rome, March 2007.

22 E.G. GUBA, Y.S. LINCOLN, *Competing paradigms in qualitative research*, in N.K. DENZIN, Y.S. LINCOLN, *The landscape of qualitative research*, Sage publications, California, 1998.

23 This is a neo-Schumpeterian notion. Joseph Schumpeter noted the importance of long-run trends (J. SCHUMPETER, *Business cycles: a theoretical, historical and statistical analysis of the capitalist process*, Mc Graw Hill Book Company, New York, 1939). "Creative destruction" (new innovations make old ones obsolete) leads to waves of innovative activity. As proprietary knowledge is protected by patents, there is an incentive to create new knowledge to reap these rewards. Schumpeter no-

Mais le rôle d'une forme d'énergie particulière en tant qu'élément central n'est pas discernable en tant que facteur de progrès de productivité et de développement de nouveaux produits, contrairement aux vagues précédentes. Le rôle de l'énergie apparaît plutôt en creux devant la nécessité sociale de corriger les effets environnementaux de ses usages et de trouver une réponse prochaine aux contraintes de ressources épuisables et au risque climatique pour assurer la soutenabilité du développement[24].

Some authors, like Réné Kemp, sees the emergence of a "Green paradigm" (new relations between the society and the nature) that could be a determinant of this new long wave[25].

J.M. Chevalier also reveals this social environmental consciousness but points out the difficulties to predict its tempo as the causalities are so different.

Ce phénomène de prise de conscience des problèmes d'environnement paraît inscrit dans l'histoire de l'humanité mais il se fait à un rythme difficilement prévisible. En effet, la socialisation de cette prise de conscience dépend de facteurs de nature différente: arguments scientifiques, souvent controversés, enthousiasmes technicistes ou écologiques, discours politiques marqués d'électoralisme, nationalisme, mondialisme, altermondialisme ou régionalisme, perception intuitive de l'avenir, avec les peurs et les espoirs qui y sont associés[26].

According to D. Finon, this new macro-social and economic paradigm should be integrated in energy modelling. A qualitative modelling of this paradigm should be translated into quantitative assumptions (cf. Shell scenarios).

Several key assumptions concerning GDP, population, oil price, and, more recently carbon price, are very similar in different energy *Bottom-up* modelling exercises like the IEA World Energy Outlook (WEM

---

ticed that periods of great innovation occurred in waves: early 1800's with water power, textiles and iron (60 years); late 1800s with steam, rail and steel (55 years); early 1900s with electricity, chemicals, internal combustion engine (50 years). One can say that these waves continue – late 1900s with petrochemicals, electronics and aviation (40 years) and from the end of the 1990s with information and communication technologies (digital networks, software and new media), biotechnologies and new materials. These waves tend to get shorter and shorter also because since the early 20th century, governments and companies intensively search for new technologies.

24 D. FINON, *Prospective énergétique et modélisation – Identification de pistes de progression méthodologique*, *op. cit.*

25 R. KEMP, *Technology and transition to environmental sustainability: the problem of technological regime shifts*, Futures, Vol. 26, 1994. The idea of "paradigm shift" was first introduced in the sociology of knowledge by Th. KUHN, *The structure of scientific revolutions*, University of Chicago Press, 1962.

26 J.M. CHEVALIER, *Les grandes batailles de l'énergie*, *op. cit.*

model), the DOE/EIA International Energy Outlook (NEMS/SAGE model), the EC World Energy Technology Outlook (POLES model), the EC scenario for the European Union (PRIMES model), the IIASA/WEC scenarios (MESSAGE model), or the EREC/Greenpeace Energy (r)evolution report (see Chapter VII)[27].

The quality of these exercises depends on the quality of assumptions on several exogenous parameters. The comment of D. Finon is apposite.

> La qualité de ces exercices de type tendanciel dépend de la qualité des hypothèses faites sur les paramètres exogènes en particulier sur la croissance économique et son contenu, sur les hypothèses de comportements (évolution sociologique) et d'institutions (niveau de libéralisme et type de mondialisation), sur le progrès techniques et les contraintes sur la diffusion d'innovations (modalité d'insertion dans les systèmes techniques existants. Elle dépend aussi de la représentation des mécanismes de marchés énergétiques (pétrole, gaz, électricité) qui reste exogène ou rudimentaire[28].

The implicit assumption – often included in the models – that the best available technologies are simultaneously being adopted in both industrial and in developing countries can be discussed. This assumption means that the political will – to accept principles like the respect of the environment, the phasing out of dirty/unsustainable energy sources, the implementation of renewables, the decoupling of economic growth from the consumption of fossil fuels or the greater equity in the use of resources – exists everywhere and that developing countries have besides the will, the means to implement those principles.

If one accept this assumption, many "heroic" or "over-optimistic" scenarios become plausible. In this respect, one can ask about the expected growth rate of developing countries in the scenarios developed by the Intergovernmental Panel on Climate Change (IPCC) where, by 2100, some developing countries become richer than OECD countries[29].

---

27 International Energy Agency, *World Energy Outlook*, *op. cit.*; US Department of Energy (EIA), *International Energy Outlook*, *op. cit.*; European Commission, *World energy, technology and climate policy outlook (WETO-2030) and World Energy Technology Outlook-2050 (WETO-H2)*, *op. cit.*; European Commission, *Energy and transport trends to 2030*, *op. cit.*; World Energy Council (WEC) and International Institute for Applied for Systems Analysis (IIASA), *Global energy perspectives*, IIASA, Laxenburg, 1998; European Renewable Energy Council (EREC)/ GREENPEACE, *Energy (R)evolution – a sustainable world energy outlook*, EREC, The Netherlands, 2007.

28 D. FINON, *Prospective énergétique et modélisation – Identification de pistes de progression méthodologique*, *op. cit.*

29 IPCC, *Special report on emissions scenarios*, *op. cit.*

Exogenous parameters

Energy outlook exercises consider four main exogenous parameters[30]:

- Economic development processes such as economic and demographic growth, land-use, industrial and services growth, degree of liberalisation, etc.;
- Constraints in terms of energy resources and in terms of climate change policies;
- Technological progress: efficiency improvements, entry date in the market of such technology, etc. (see next part of this Chapter on climate change and technological progress);
- Behaviours, life styles and public policies (*variables d'action*).

Driving forces, based on past and present trends, are the underlying forces that determine the way things happen, setting the pattern of events and determining outcomes[31]. Understanding them is a key function of scenario development. Pierre Wack from Shell spoke about "predetermined events" – like demographic change – where the momentum of driving forces already in place made future outcomes highly probable[32].

As a complement to driving forces, the *Global Scenario Group* introduced the concept of "attractive and repulsive forces" that are defined as the forces that can substantially redirect beliefs, behaviour, policies and institutions towards some futures and away from others[33].

D. Finon speaks about the regularity of the tendencies and the game of compensations. A change in a sub-system implies adaptations in other sub-systems. In a Bottom-up detailed representation of technologies and driving forces, if one can see a decoupling between energy growth and economic growth, this is not always the case as demonstrated by the electrical use and by the transport sector[34]. This would mean that there would be certain constancy in the "energy budget".

For example, if energy efficiency in buildings are implemented (better insulation, double-glazing, etc.), the consumers can choose to in-

---

30 D. FINON, *Prospective énergétique et modélisation – Identification de pistes de progression méthodologique*, *op. cit.*

31 K. VAN DER HEIJDEN, P. SCHUTTE, *Key questions for designing scenario applications*, *op. cit.*

32 P. WACK, *Scenarios: Uncharted waters ahead*, *op. cit.*

33 G. GALLOPIN, A. HAMMOND, P. RASKIN, R. SWART, *Branch points: Global scenarios and human choice*, Stockholm Environment Institute, Scenario Group, SEI PoleStar Series Report No. 7, Sweden, 1997.

34 World Energy Council, *Drivers of the energy scene*, WEC, London, 2002.

crease the ambient temperature of their home. In the transport sector, the rules like the ones developed in the Clean Air for Europe-CAFÉ or in the Voluntary Agreements between the European Commission and the Association of European Car Manufacturers-ACEA obliging automakers to increase the efficiency of their vehicles can contribute to the tendency of consumers to buy a car of a superior category[35].

De facto, these "invariants" in the vocabulary of D. Finon are only valid at the aggregate level.

> Ces invariants ne se retrouvent pas au niveau désagrégé des pays et des classes de consommateurs (...). Des recherches sur ces invariants permettraient de positionner les résultats de certains scénarios (...) et de comprendre les interactions entre sous-systèmes à la base de ces effets de compensation, car elles ne peuvent qu'évoluer dans la longue durée (...). La construction de scénarios normatifs doit se baser sur la "déconstruction" de tendances pour comprendre les dynamiques à l'œuvre[36].

"Reference"

The Reference year is the year from which starts the impact simulation exercise. The Reference case (Business as Usual) includes the energy supply and demand trends taking into account the main energy drivers (GDP, population, households, industry and transport growth rates, etc.). The policy scenario is the energy supply and demand developments taking into account additional energy-policy related measures (like Kyoto targets, energy taxation, energy savings measures, etc).

At the EU-level, the Heads of State and Governments of the twenty-seven Member States fixed ambitious quantified targets on greenhouse gas emissions, on energy efficiency and on renewable energy sources by 2020[37]. Sometimes, the comparison is made to a Reference year, to a Reference case (scenario or projection) or to a specific percentage. As a concrete example, the March 2007 European Council committed the EU to:

- Achieve at least a 20% reduction of greenhouse gas emissions by 2020 *compared to 1990*;

---

35 European Commission, *Clean Air for Europe (CAFE) Programme*, COM(2001)245; European Commission, *Thematic strategy on air pollution*, COM(2005)446; European Commission, *Implementing the Community strategy – Reduce CO_2 emissions from cars: An environmental agreement with the European automobile industry*, COM(98)495.

36 D. FINON, *Prospective énergétique et modélisation – Identification de pistes de progression méthodologique*, *op. cit.*

37 Council of the European Union, *Presidency conclusions*, Brussels European Council, 8-9 March 2007.

- Increase energy efficiency in the EU so as to achieve the objective of saving 20% of the EU energy consumption *compared to projections for 2020*;
- The binding targets of *20% share* of renewable energies in overall EU energy consumption by 2020 and of 10% binding minimum target to be achieved by all Member States for the share of biofuels in overall EU transport petrol and diesel consumption by 2020.

In another way, some authors criticize the small ambition between a Reference case and a Policy scenario. For example, according to Bernard Laponche, in the 2006 *World Energy Outlook* of the International Energy Agency, by 2030, the difference in global energy consumption is only 10% between the "Reference scenario" and the "Clean and Clever scenario". In this latter scenario, fossil fuels remain very high and the energy efficiency improves by only 7% in 25 years. In addition, the industrialised countries consumption *pro capita* do not really change[38].

To these remarks, Fatih Birol (responsible of the WEO of the IEA), answers that the Reference scenario is based on national policies already adopted. The Clean and Clever scenario is based on energy policies, regulations and incentives – essentially focusing on energy security of supply and greenhouse gas emission reductions – that are currently discussed, but not yet implemented. For F. Birol, these scenarios are realistic. They aim to assess the impacts of existing policies in the Reference scenario and to assess the impacts of more ambitious policies – currently discussed – in the Clean and Clever scenario.

In the Reference case of the European Commission WETO-H2 study, one can see that the place of Europe is strongly being reduced in terms of energy consumption (cf. Figure IV.3)[39]. If Europe accounted for more than 20% of energy consumption in 1990, it account for 15% in 2010 and probably less than 10% in 2050.

---

38 B. LAPONCHE, F. BIROL, *Le World Energy Outlook 2006 est-il timoré*, Enerpresse, 9250, Paris, 26/01/2007.

39 European Commission, *World Energy Technology Outlook-2050 (WETO-H2)*, *op. cit.*

Figure IV.3: World primary energy consumption and the place of Europe in a Reference scenario

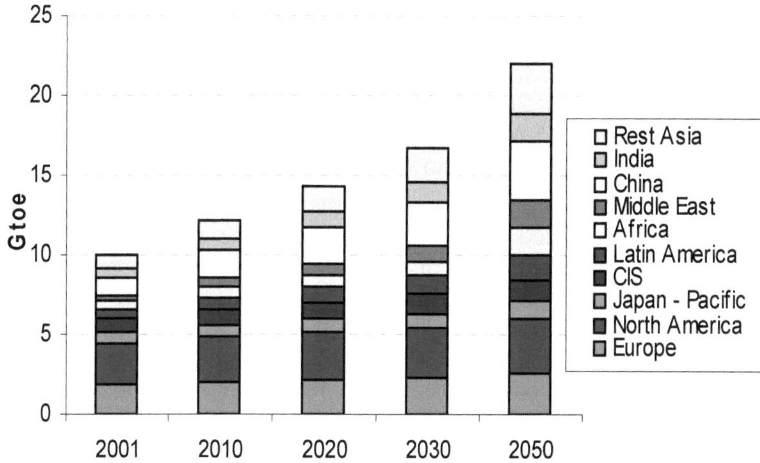

Discount rate

For some aspects, like climate change, where 100 years or more are considered, the discount rate used is crucial (e.g. 0%, 3%, 6%, 8% or 10%). The discount rate allows to treat the intergenerational equity. More the rate is close to 0, more the weight given to the welfare of future generation is important (there is no difference between the welfare of future generations and the welfare of present generations). More this discount rate is high, more it reflects the preference of the society for the present.

The Stern Review calls for:

> prompt and strong action… If we don't act, the overall costs and risks of climate change will be equivalent to losing at least 5% of global GDP each year, now and forever. If a wider range of risks and impacts is taken into account, the estimates of damage could rise to 20% of GDP or more.[40]

William Nordhaus finds that optimal economic policy for climate change involves modest rates of emissions reductions in the near term: 6% emissions reduction in 2005; 14% in 2050; and 25% in 2100. According to Nordhaus, the Stern Review results are "dramatically different" from existing economic analyses[41].

---

40 N. STERN, *Review of economics of climate change (Stern Review)*, Report commissioned by the UK Chancellor of the Exchequer, London, November 2006.

41 W. NORDHAUS, *Critical assumptions in the Stern Review on climate change*, Science, Vol. 317, 13 July 2007.

Two important differences are that Stern assesses damages through 2200, and uses a discount rate that varies around a central value of 1.4%, while Nordhaus assesses damages through 2100 and uses a discount rate that varies around a central value of about 4%.

With a discount rate of 4%, € 1 at 100 years from now is worth € 0.018 today; € 1 at 200 years from now is worth € 0.0003 today. With a discount rate of 1.4%, € 1 at 100 years from now is worth € 0.25 today; € 1 at 200 years from now is worth € 0.06 today.

A first and important difference is therefore coming from the chosen time frame and discount rate. Nevertheless, the difference is not just discounting: the assessment of future damages is much greater in the Stern Review using the PAGE model (some categories of damages are omitted or under-estimated in the Nordhaus DICE model in comparison to the Stern Review like the impacts on agriculture, water or health). The assessment of costs of emission reduction is lower in the Stern Review than in Nordhaus analyses. In addition, concerning the treatment of uncertainty, the Stern Review includes allowance for risk aversion, i.e. it takes the view that climate policy is partly a matter of insurance

Past results

A study made in the US at the end of the 1970s was already critical about the quality of modelling results.

> The major studies since 1973 have given us predictions about the US energy situation that have consistently been more optimistic than the reality has proved to be, especially in regard to energy supplies. Some of the models were published without receiving much notice and had virtually no lasting impact. Nevertheless, it seems abundantly clear that some of the optimistic forecasts issued did influence – and mislead – both the energy policymakers and the informed public about the causes and possible solutions for the energy problem[42].

Some projections have been close to reality. One European example for the total energy consumption will be analysed in Chapter VII (exercise made in 1985 with the EFOM and MEDEE models for the European Community in 2000).

Some predictions and warnings were very close to reality. This was the case, for example with the claim that continuing release of chlorofluorocarbons (CFC) to the atmosphere would reduce concentrations of stratospheric ozone proved essentially correct[43].

---

[42] R. STOBAUGH, D. YERGIN (eds.), *Energy Future*, Random House, New York, 1979.

[43] M.J. MOLINA, R.F. ROWLAND, *Stratospheric ozone sink for chlorofluoromethanes: chlorine atom catalyzed destruction of ozone*, Nature, Vol. 249, 1974.

Another successful projection study in the US was led by Sam Schurr and Bruce Netschert in 1960. They used a combination of extrapolation from historical trends and common sense concerning saturation of energy demand. They arrived at an aggregate projection for the US for 1975 of 75 quads to be compared with the real US energy demand that was 71 quads in that year[44].

The preciseness of past modelling results has a clear influence on the credibility of the today projections. As Paul Zagamé said :

> La précision des prévisions passées rejaillit sur l'audience des prévisions d'aujourd'hui[45].

Sociological aspects

Several sociologists concluded that rather than influencing the socio-political system, models are influenced by the socio-political system. In the spirit of Bruno Latour, as in the case of any other scientific or technical object, the construction of a model is the result of collective process (*foi dans les modèles*)[46].

Martin Greenberger, Matthew Crenson and Brian Crissey highlighted another issue:

> Because of their flexibility, models can do more than provide a framework for debate. They can become the instrument of debate[47].

Craig Galbraith and Gregory Merrill wrote "models for forecasting are socially used and produced[48]". In his analysis and comparison of the use of energy modelling in UK and in The Netherlands, Reiner De Man concluded:

> The nature of the political context is a far more significant explaining factor (for the ways in which energy scenarios are built) than the substance of the models themselves[49].

---

44 S. SCHURR, B. NETSCHERT, *Energy in the American economy*, 1859-1975, John Hopkins Press, Baltimore, 1960.

45 P. ZAGAME, *L'expérience française de modélisation macro-économétrique: bilan et perspectives*, *op. cit.*

46 LATOUR B., *Science in action*, Harvard University Press, 1987.

47 M. GREENBERGER, M. CRENSON, B. CRISSEY, *Models in the policy process – Public decision making in the computer area*, Russell Sage Foundation, New York, 1976.

48 C.S. GALBRAITH, G.B. MERRILL, *The politics of forecasting – Managing the truth*, California Management Review, Vol. 38, 1996.

49 R. DE MAN, *Energy forecasting and the organization of the policy process*, Eburon, Delft, 1987.

In other words, the construction of energy models is bound to certain times and places, to socio-political context. As Bastiaan de Laat wrote:

> The construction of the model is an interactive work between labs, institutes, firms, government departments, etc. A model is the result of a trajectory or a sequence of decision on what the model should contain. Finally, a model follows a specific socio-technical script with the type of actors involved and the reasons for which the model has been developed[50].

This question of credibility of the authors is also particularly sensitive in the energy field. For example, serious studies made by the International Atomic Energy Agency could even not be open by ecologists and NGO while serious studies made by Greenpeace could never arrive on the fossil fuels and nuclear experts tables.

In the same line, several authors highlighted the following questions related to modelling: What is the issue? Who is the client? Whose project is it? What is the question?[51]

Already in 1979, Alan Manne underlined what one could call the "administrative risk of modellers":

> During the years since the Arab oil embargo of 1973, energy modellers have been in the position of those ancient messengers who had the unenviable task of bringing bad news to their monarchs. Today, modellers are unlikely to suffer decapitation, but still – those who bring bad news are unlikely to enjoy an increase in their computer budgets and research staffs (…). To protect the research organization, it is much safer to construct a complex model that is never quite ready to run – rather than a simple analysis which produces commonsense answers[52].

Speaking about the *modèle de la rationalité technique* (technical rationality model), Philippe Roqueplo highlights the links between the experts and the deciders. But he adds that

---

50 B. DE LAAT, *Sociology for energy modelling and the sociology of energy models*, in GIOVANNINI B., BARANZANI A., *Energy modelling beyond economics and technology*, CUEPE, Genève, 1997.

51 A clearly liberal politically-oriented researcher, Dan Stastny concluded his paper by these points: Politicians are interested in conducting as much policy as possible and their policy should be justified in the eyes of public; Economics can provide politics with this justification. Politics exert influence over economic science, for example by identifying "market failures" to become less anti-interventionist and more pro-interventionist. D. STASTNY, *Economics and politics: Mutual relationship*, Working Paper, Austrian Scholars Conference, 15-16 March 2002.

52 A.S. MANNE, *The use and role of models: Long-term energy projections for the USA*, International Conference of Dublin, *Energy systems analysis*, 9-11/10/1979, D. Reidel-Kluwer, ECSC-EEC-EAEC, Brussels and Luxemburg, 1980.

These two ways of deciding are not sufficient and an equilibrium should be found among the three sides of the triangle, and therefore also between the experts and the citizens[53].

A last sociological element is the predisposition or not to take economic analysis. For example, in the energy and environmental field, there is a fertile background to use quantitative tools also due to international, European and *ad hoc* networks like the International Association of Energy Economists, the European Network for Energy Economics Research and the ENVECO Group composed of economists from Members States and the Commission from the Environmental and the Economic and Finance departments.

Technical and geographical representation

According to D. Finon, in the Bottom-up exercises until 2030, three main improvements could be made:

- A better representation of the potential actions;
- A better representation of the potential dynamics of some specific technical systems like hydrogen, decentralised electricity production or carbon capture and storage;
- A better representation of options and constraints in emerging countries like China, India and Brazil.

Speaking about the development of new energy technologies, Bernard Bourgeois, Dominique Finon and Jean-Marie Martin underline the all or nothing effects (*effet de tout ou rien*) in the linear models and the low reliability of the hypotheses about the future cost of different technologies, as only two methods exist: the extrapolation of past trends and the expert opinions[54]. D. Finon has developed this observation.

> L'effort d'amélioration doit porter sur la représentation des marges de manœuvre d'ici 2020-2030 et des dynamiques possibles de développement des systèmes techniques particuliers: les hydrocarbures de synthèse, les systèmes hydrogène, la production électrique décentralisée, les modes de séquestration du carbone. De même en est-il de l'analyse des contraintes et des options se présentant aux grands pays du Sud (Chine, Inde notamment) qui sont appelés à peser de plus en plus lourd dans l'équilibre des marchés énergétiques mondiaux, le prélèvement des ressources mondiales et les émissions de CO_2[55].

---

53 Ph. ROQUEPLO, *Penser la technique*, Seuil, Paris, 1983; Ph. ROQUEPLO, *Climats sous surveillance – Limites et conditions de l'expert*, Paris, 1993.

54 B. BOURGEOIS, D. FINON, J.M. MARTIN, *Énergie et changement technologique – une approche évolutionniste*, Economica, Paris, 2000.

55 D. FINON, *Prospective énergétique et modélisation – Identification de pistes de progression méthodologique*, *op. cit.*

For example, as suggested in one of the Shell scenarios, China could follow an original pathway based on coal[56].

Figure IV.4: Daily time spent on food vs. information – in 2000 and 2100

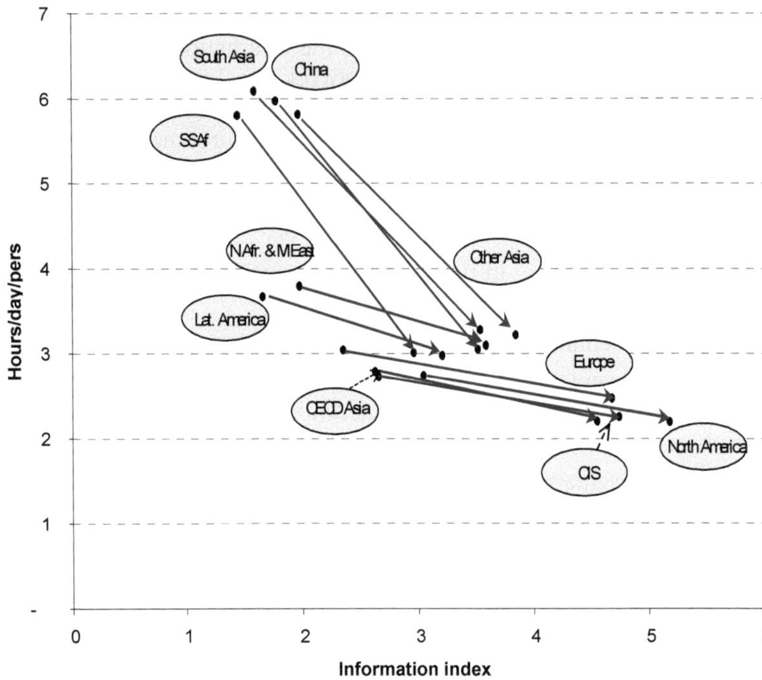

Bertand Château, Thomas Hamacher, Koen Smekens, Dag Martinsen, Hubert Reisinger, Martin Patel underlined the tendency towards less and less time spent for food reasons (agriculture, meal preparation, etc) *versus* more and more time spent on education and on gathering general knowledge ("Information Index")[57].

---

56 Shell, *Energy needs, choice and possibilities: scenarios to 2050*, *op. cit.*

57 B. CHATEAU, T. HAMACHER, K. SMEKENS, D. MARTINSEN, H. REISINGER, M. PATEL, *Very long term energy modelling (VLEEM)*, Final report, European Commission, Brussels, 2000.

Long-term and "Back-casting"

In different modelling studies (IEA, European Commission or US DoE/EIA), at a 20-30 years horizon time, unless very strong policy measures are introduced, similar messages emerge: almost three quarters of the world's energy supply continue to come from fossil fuels; energy demand and CO_2 emissions approximately double; electricity consumption increase four-fold. No major technological or behavioural breakthroughs are envisaged that could make a substantial change in the energy system[58].

Several experts speak about the challenges of the long-term and the difficulties to take into consideration the risks, opportunities, constraints, innovation trajectories, behaviours and availability of resources in such a time-frame. Jean-Marie Martin wrote:

> Les décisions énergétiques prises sur l'horizon 2010-2020 ont de fortes chances d'être entachées de myopie car incapables de prendre en compte les risques, opportunités et contraintes à beaucoup plus long terme[59].

Some authors like Gill Ringland even speak about the "tyranny of the present"[60].

D. Finon highlighted the possibility of having a completely different technological and social system after 2030.

> Le dépassement de l'horizon 2030 induit une vision différente des marges de manœuvre, des trajectoires d'innovations et de comportements, et du durcissement des contraintes sur les ressources épuisables et la précaution climatique (...). L'affranchissement progressif de l'héritage du système technologique en place au fur et à mesure du renouvellement des équipements énergétiques, industriels et des infrastructures (logement, transports) conduit à envisager des ensembles technologiques totalement renouvelés, que ce soit du côté de l'offre que des usages de l'énergie. Il invite aussi à anticiper les réponses des firmes énergétiques à la raréfaction très progressive des ressources. De la même façon l'éloignement de l'horizon permet d'imaginer une évolution radicale de la relation des sociétés à l'environnement naturel[61].

In the same line and from a socio-cultural perspective, Bernard Giovannini and Andrea Baranzini wrote:

---

[58] International Energy Agency, *World Energy Outlook*, *op. cit.*; US Department of Energy (EIA), *International Energy Outlook*, *op. cit.*; European Commission, *World energy, technology and climate policy outlook (WETO-2030) and World Energy Technology Outlook-2050 (WETO-H2)*, *op. cit.*

[59] B. BOURGEOIS, D. FINON, J.M. MARTIN, *Énergie et changement technologique – une approche évolutionniste*, Economica, *op. cit.*

[60] G. RINGLAND, *Scenarios in business*, John Wiley & Sons, Chichester, 2002.

[61] D. FINON, *Prospective énergétique et modélisation – Identification de pistes de progression méthodologique*, *op. cit.*

> In the long run, everything may change, from the economy production structure to consumer habits and values (...). If in the short – to medium terms (0-10 years), modelling can safely consider that socio-economic and technological factors are fixed (...), in the long-run (10-100 years), structural, societal and cultural changes (or even revolution) should be taken into account[62].

Douglas North adds the institutional dimension "For the long-term issues it is fundamental to understand how and why cultural and institutional factors evolve and how they can be related to human behaviour in general and energy consumption in particular"[63].

Concerning transport for example, if one counts that any long distance flight now produces at least 1 tonne of CO_2, and that some very efficient scenario in 2050 speak about a per capita allocation of 1.3 tonne per year in industrialised countries, the (r)evolution envisaged in some scenario (like the EREC/GREENPEACE scenario) will have to be mostly in behaviours and social value[64].

Speaking about the complexity of the energy and environmental problems, J.M. Martin insisted on the precaution principle and on the technological choices that should respect the intergenerational welfare and the cross-border negative externalities.

> La recherche de solutions est rendue difficile par la complexité des problèmes et le caractère lointain, incertain, parfois planétaire des risques. Un accord pourrait néanmoins se dégager autour du principe de prudence qui préconise de repousser la mise en œuvre d'une technologie risquée jusqu'à l'obtention d'une information scientifique et technique plus complète. Son application incombe aux États dont les missions de préservation de l'intérêt collectif s'élargissent désormais dans une double direction: temporelle, avec la prise en charge du bien-être intergénérationnel; spatiale, avec le traitement des externalités négatives transfrontalières. Ce nouveau contexte institutionnel devrait s'imposer aux firmes, notamment dans leurs choix technologiques[65].

Criticizing traditional Top-down and Bottom-up models for dealing with long-term and complex global problems, Pascale Morand Francis said:

---

62 B. GIOVANNINI, A. BARANZANI, *Energy modelling beyond economics and technology*, CUEPE, Geneva, 1997.

63 D.C. NORTH, *Institutions, institutional change and economic performance*, Cambridge University Press, UK, 1990.

64 EREC/GREENPEACE, *Energy (r)evolution – a sustainable world energy outlook*, *op. cit.*

65 B. BOURGEOIS, D. FINON, J.M. MARTIN, *Énergie et changement technologique – une approche évolutionniste*, *op. cit.*

Neoclassical concepts, designed for short-term analysis, presume permanence rather than process, and are based on the erroneous assumption that it is possible to predict macroeconomics from microeconomics, the global economy from national economics, and societal behaviour from individual behaviour[66].

Another major issue highlighted since several decades concerns the difference between natural and social sciences. As Walter Elsasser pointed out in 1960, the immense power of mathematics in physics derives from the homogeneity of the classes that physics deals with. All hydrogen atoms are identical, all electrons are identical, all pi-mesons are identical. By contrast, as one moves to biology and then into the social sciences, the individuals under study no longer are identical – indeed variability of individuals becomes an essential attribute. Such variability tends to defy categorization and treatment by mathematics[67].

In addition, as mentioned by Alvin Weinberg,

> the phenomena of the social sciences tend to be what I call "rare" events: each business cycle, or political upheaval, or social change, is unique, i.e. rare. And mathematical modelling lacks proficiency when it tries to capture "rare" events[68].

The more the time-frame expands, the more uncertainties increase. This is true in terms of economic and technological development, social behaviours and availability of resources. Therefore, to get coherent images of the far future become more and more difficult. One possible answer to the very-long term issues suggested by D. Finon is closely connected to the Back-casting approach.

> Le problème est à traiter par une démarche en deux étapes: Premièrement, définir une image normative de long terme selon un critère de soutenabilité en vérifiant sa compatibilité avec les contraintes d'équilibre général; cette première étape implique de se référer aux contraintes de ressources et de réaliser des maquettes macro-économiques mondiales articulant variables démographiques et dynamiques de croissance, d'échanges de capitaux et de flux de marchandises et d'énergie; Deuxièmement, tracer à rebours une trajectoire de transition en compatibilité avec les contraintes économiques, les inerties associées au stock de capital technique et aux comportements, et les effets d'irréversibilité en s'assurant également des cohérences technologiques, sociales et géopolitiques intrinsèques au scénario (...). La modélisa-

---

66 P. MORAND FRANCIS in B. GIOVANNINI, A. BARANZANI, *Energy modelling beyond economics and technology*, CUEPE, Geneva, 1997.

67 W. ELSASSER, *Atom and organism: A new approach to theoretical biology*, Princeton University Press, 1966.

68 A. WEINBERG, *Limits to energy modelling*, International Conference of Dublin, *Energy systems analysis*, 9-11/10/1979, D. Reidel-Kluwer, ECSC-EEC-EAEC, Brussels and Luxemburg, 1980.

> tion doit permettre de tracer des trajectoires cohérentes qui intègrent les effets d'irréversibilité des choix successifs sur les variables d'action, notamment en prenant en compte les dynamiques incorporées par l'endogénéisation du progrès technique[69].

The European research project VLEEM funded by the 5th RTD Framework Programme (See Chapter II) developed a Very Long Term Energy-Environment Model essentially dealing with climate change and depletion of fossil fuel resources[70]. VLEEM combined the Back-casting method and the re-think of energy-environment modelling to assess the relationships between very long-term changes in social and cultural behaviour and technological progress. Human activities were categorised into five main social and cultural areas: economic production, food and feeding, shelter and lodging, leisure and self-fulfilment, and mobility. The energy and services needed for each of these activities were then projected into the future according to different scenarios.

Population is the fundamental driver of the needs for energy services over the very long-term. As well as the number of persons, the number of households also has an important influence on energy needs. The average number of persons per household in the world is expected to decrease from 3.7 in 2000 to 2.4 in 2100, with decreases by more than half in Africa, the Middle East and South Asia.

Information (education and general knowledge taken from newspapers, TV, Internet, travels, etc.) is the second key driver of energy needs. It determines wealth through labour productivity, and hence consumer behaviour in terms of quantities and technical specifications of goods and services demanded.

How people divide their time between work and leisure is the third main driver. Household income depends on the time spent in working for money that is then spent on the purchase of goods and services, thereby increasing energy needs. The needs for energy services diversify and increase along with the decrease in time spent per activity. For instance, eating a ready-made frozen dish saves a lot of cooking time, but results in additional energy consumption in the home for transporting, freezing, storing and warming up the meal.

The five main conclusions that can be drawn for the changes in worldwide energy needs from 2000 to 2100 are the following in a

---

69 D. FINON, *Prospective énergétique et modélisation – Identification de pistes de progression méthodologique*, *op. cit.*

70 The VLEEM project was coordinated by ENERDATA (Bertrand Chateau) with the participation of Max Planck-IPP (Thomas Hamacher), ECN (Koen Smekens), FZ-JULICH (Dag Martinsen), VERBUND (Hubert Reisinger) and University of Utrecht (Martin Patel).

"business as usual" scenario (i.e. according to the social and economic patterns related to the expected socio-economic development):

- Energy used for food will double;
- Energy for thermal comfort and living-space heating and cooling will triple;
- Energy for mobility will increase by a factor of five;
- Energy for economic production will only increase moderately;
- Energy for leisure and self-fulfilment (sports, cultural, social, and tourism activities) will have the fastest growth of all.

Figure IV.5: Number of persons per household (2000-2100)

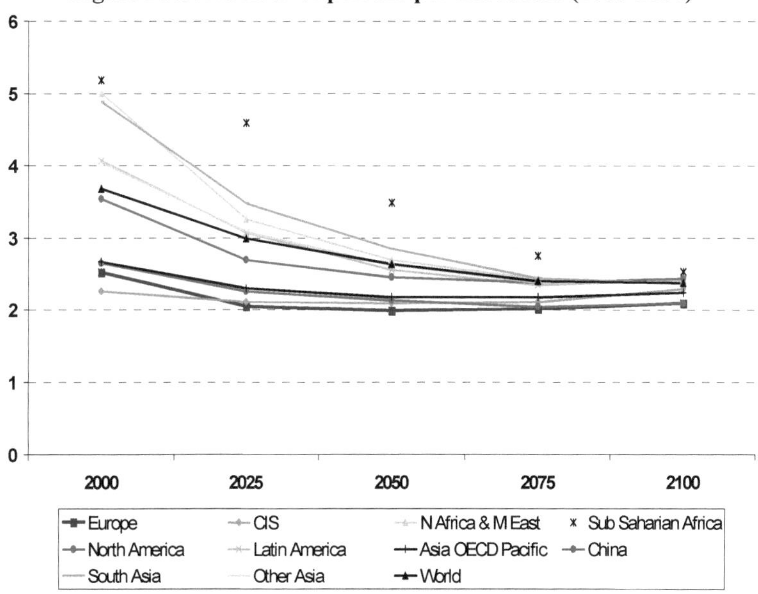

Technological progress and climate change policies

The relation between R&D, technological innovation and economic performance has been studied in details since the 1990s[71]. Technological innovation is seen as the major source of the productivity gains that

[71] P.M. ROMER, *Endogenous technological change*, Journal of Political Economy, No. 98, 1990; D. BLONDEL, *L'innovation pour le meilleur et pour le pire*, Hatier, Paris, 1990; G.M. GROSSMAN, E. HELPMAN, *Innovation and growth in the global economy*, Cambridge, MIT Press, 1991; P. AGHION, P. HOWITT, *Endogenous growth theory*, MIT Press, Cambridge MA, 1998.

characterise the New Economy[72]. According to Dale Jorgenson, the New Economy refers to a fundamental transformation in the industrialised countries as businesses and individuals capitalise on new technologies, new opportunities and new national investments in computing, information and communication technologies[73]. Already in 1980, Alvin and Heidi Toffler spoke about the extra-intelligent networks that revolutionise the whole economy and society[74].

Jean-Hervé Lorenzi and Jean Bourlès highlight the today complexity of the inter-sectoral relations that makes difficult to find out the new major innovation that will give birth to a new Kondratiev cycle:

> La complexification des relations intersectorielles explique en partie la difficulté de sélectionner l'innovation majeure qui va jouer un rôle moteur (...). Le critère qui permettra de déterminer qu'une innovation est majeure ne peut être que la capacité du secteur auquel elle appartient à mobiliser et à transformer – par l'innovation – le maximum de secteurs, en termes de poids dans l'économie, pendant une période compatible avec la durée de vie de cette innovation[75].

In the energy field, according to Nebosja Nakicenovic, Arnulf Grubler and Alan Mac Donald, technological change and economic growth are the major determinants of the future global energy demand, CO_2 emissions and climate impacts. The compatibility between the increasing world energy needs and the protection of the environment will depend from the change in energy technologies. The scenarios built by various organisations like the IEA, the WEC, the EC or the IIASA illustrate this prominence of technological change from both the supply and the demand side. N. Nakicenovic *et al.* wrote:

> The evolution of technological systems is as important as the evolution of individual new technologies. As the use of individual technology, such as the automobile, expands, its evolution becomes interwined with the evolution of many other technologies and institutional and social developments[76].

The IIASA researchers were among the first, in 1995, to include technological progress as an endogenous process in a Bottom-up model (MESSAGE). They noticed that learning-by-doing in the energy sector

---

72 OECD, *Is there a new economy*, a first report on the OECD Growth project, OECD, Paris, 2000.

73 D. JORGENSON, *Deconstructing the computer, measuring and sustaining the new economy*, The National Academies Press, Washington D.C., 2005.

74 A. TOFFLER, H. TOFFLER, *The third wave*, Bantam books, New York, 1980.

75 J.-H. LORENZI, J. BOURLES, *Le choc du progrès technique*, Economica, Paris, 1995.

76 N. NAKICENOVIC, A. GRUBLER, A. MAC DONALD, *Global energy perspectives*, Cambridge University Press, 1998.

could revolutionise the system in the 21st century to such a degree that a very low carbon future could be just as cheap as a high carbon future. Such system would be heavily dependent upon the policy choices taken early in the century[77]. In the same time, Top-down modellers started to explore the incorporation of both explicit knowledge functions using "accumulated R&D" and "learning by doing" equations.

Using the NEMESIS econometric model, Arnaud Fougeyrollas, Pierre Le Mouël and Paul Zagamé have demonstrated that technological change can have large effects on emission reduction efforts to meet targets for the reduction of emissions of greenhouse gases:

> L'endogénéisation du progrès technique allège le coût de ces politiques, à la fois en termes de PIB et d'emploi[78].

The Impact Assessment of the seventh RTD Framework Programme, largely made with the NEMESIS model, also highlighted that R&D boosts economic growth and is positive in terms of employment (up to 1% of GDP growth and 1 million jobs created in 2030 in a "voluntarist" EU Research scenario)[79].

The knowledge variable and its externalities (knowledge spillovers[80]) play a major role in the endogenisation of technical change. It explains non-decreasing returns (accumulation without saturation of production factors such as R&D and human capital)[81]. In order to adapt the NEMESIS model to the new theories of growth and technical change, Carole Chevallier, Arnaud Fougeyrollas, Pierre Le Mouël and Paul Zagamé have included several phenomena in the model: the endogenization of technical change (innovation), spillovers effects (knowledge transfers) and the economic consequences of innovation.[82]

Arnaud Fougeyrollas, Pierre Le Mouël and Paul Zagamé have underlined the two reasons for which new growth theories – like endogenous

---

77 A. GRUBLER, N. NAKICENOVIC, D.G. VICTOR, *Dynamics of energy technologies and global change*, Energy Policy, Vol. 27, 1999.

78 A. FOUGEYROLLAS, P. LE MOUEL, P. ZAGAME, *Les nouvelles théories de la croissance en application: l'évaluation des politiques structurelles – Le cas du protocole de Kyoto*, Revue Économique, Vol. 56, No. 5, Paris, Septembre 2005.

79 European Commission, Staff working document, *Annex to the proposal for the Council and European Parliament decisions on the seventh EU RTD Framework Programmes* (EC and Euratom), COM(2005)119 and SEC(2005)430.

80 Z. GRILICHES, *Issues in assessing the contribution of Research and Development to productivity growth*, Bell Journal of Economics, 1979.

81 P. AGHION, P. HOWITT, *Endogenous growth theory*, *op. cit.*; P.M. ROMER, *Endogenous technical change*, *op. cit.*

82 C. CHEVALLIER, A. FOUGEYROLLAS, P. LE MOUEL, P. ZAGAME, *A time to sow, a time to reap for the European countries: a macro-econometric glance at the RTD national action plans*, Revue OFCE, June 2006.

growth – have not been included in the Top-down models. It could be because new complex theories are difficult to translate in large models; or it could be because statistics are missing to achieve reliable calibrations or strong econometrics estimations.

> L'écart entre les avancées théoriques et l'état de l'art de la modélisation appliqué est aujourd'hui grandissant: de nombreuses théories récentes n'ont pas été introduites dans les modèles opérationnels, modèles d'équilibre général ou modèles économétriques. Deux principales raisons peuvent être invoquées pour cela: les phénomènes décrits par les nouvelles théories (de la croissance) conduisent à des formulations complexes difficiles à traduire dans ces modèles souvent de très grande taille; par ailleurs, on manque souvent de statistiques pour aboutir à des calibrations fiables ou à des estimations économétriques robustes. Les nouvelles théories de la croissance sont clairement dans ce cas[83].

Looking at the impact of endogenous technical change (ETC) using the long-term PAGE2002 integrated assessment model, Stephan Alberth and Chris Hope arrived at the conclusion that:

> For most standard abatement paths there would be an initial "learning investment" required that would substantially reduce the unit costs of CO_2 abatement as compared to a business as usual scenario[84].

Therefore, induced technological progress should be looked at for an effective climate change policy.

Making the synthesis of four papers – from Georg Müller-Fürstenberger and Gunter Stephan, Richard Tol, Socrates Kypreos, and Claudia Kemfert and Truong Truong – Philippe Thalmann and Laurent Viguier pointed out that all these contributions, "find large reductions in stabilization costs from different forms of endogenous and exogenous technological change, with numerical simulations to the end of this century"[85].

The learning by doing concept (experience curve) was first developed in the 1930s by Thomas Wright in the aircraft industry[86]. He described a basic theory for obtaining cost estimates based on repetitive production of airplane assemblies: the repetition of the same operation

[83] A. FOUGEYROLLAS, P. LE MOUEL, P. ZAGAME, *Les nouvelles théories de la croissance en application – L'évaluation des politiques structurelles, le cas du protocole de Kyoto*, Revue Économique, Vol. 56, No. 5, septembre Paris, 2005.

[84] S. ALBERTH, C. HOPE, *Climate modelling with endogenous technical change: stochastic learning and optimal greenhouse gas abatement in the PAGE2002 model*, Energy Policy, Vol. 35, March 2007.

[85] Ph. THALMANN, L. VIGUIER, Editorial, *Technology in climate policy and climate models*, Energy Policy, Special edition, Vol. 35, No. 11, November 2007.

[86] T.P. WRIGHT, *Factors affecting the cost of airplanes*, Journal of Aeronautical Sciences, 3, 1936.

results in less time or effort spent on that operation. In the Wright learning curve, the underlying hypothesis is that the direct labour man-hours necessary to complete a unit of production will decrease by a constant percentage each time the production quantity is doubled. If the rate of improvement is 20% each time that quantity is doubled, then the learning percent (progress ratio) is 80% (100 - 20 = 80). In the 1960s, the Boston Consulting Group made popular the concept of "experience curve" to describe cost reduction through experiential learning.

Learning by doing as a source of technical change was firstly emphasised by Kenneth Arrow at the beginning of the 1960s[87]. Learning by doing is the acquisition of new skills, new technology or new processes that occurs during the production process. It takes place after a technological innovation has been adopted and it allows to do something better. Thesc gains are internal to the production process.

In the 1970s, Nathan Rosenberg noted that R&D involved several different forms of learning: basic research (learning about the laws of nature) and development (learning about the optimal design characteristics of a product). Learning was focused on commercial dimensions of the innovation process[88]. However, experience was also important as many product improvements were made by production workers, managers, and other users of technology. N. Rosenberg noted that some parts of learning by doing received less attention because they did not received direct expenditures. Experience (with specialised knowledge, training and skills) led to improvement.

Learning by using represents technological advances that occur during utilisation of the product (as opposed to during the production of the product). Learning by using occurs in response to new opportunities or lessons gained from experience. Learning by using creates two different types of useful knowledge:

- Embodied knowledge (early experience with a new technology leads to better understanding of the relationship between basic design characteristics and performance) and
- Disembodied knowledge (knowledge leading to changes in use of the product that do not require modifications in hardware design for example, when new practices for using the product are developed. N. Rosenberg considered this the purest form of learning by using).

---

87 K. ARROW, *The economic implications of learning by doing*, The Review of Economic Studies, Vol. 29, No. 3, June 1962.

88 N. ROSENBERG, *Perspective on technology*, Cambridge University Press, 1976; N. ROSENBERG, *Inside the black box: Technology and economics*, Cambridge University Press, 1982.

Induced innovation is an example of a "Demand-pull" influence, i.e. at the role of market prices on technological change[89] (On the contrary, "Technology push" focuses on the impact of R&D on innovation). For example, the more labour prices increase, the more efforts are made in R&D to save labour. The model starts with an innovation possibility curve that tells all the possible techniques that could be used at a given time. Before any one technique is chosen, R&D must be performed[90]. The technology chosen depends on the relative prices. A clear example related to energy prices happened at the end of the 1970s. When energy prices were the highest, patents related to new energy sources peaked.

In its critic of the "one side focus" of J. Schmookler, N. Rosenberg underlined that the demand-pull can only be based on the technological competences of firms. These competences do not always favour the deployment of the most advanced and clean technologies[91]. Therefore, the question arises as to how to promote new and clean energy tech-

---

89 Demand-pull theories focus on the benefits arising from a new innovation. They focus on product, rather than process. The demand-induced hypothesis tested by Jacob Schmookler was that greater market demand led to increases in innovation within industries. Schmookler's results indicated a positive correlation between added-value and the number of patents supporting the demand-induced hypothesis. J. SCHMOOKLER, *Invention and economic growth*, Harvard University Press, 1966). Evolutionary theory is different from induced innovation. For the Evolutionary theory – term borrowed from biology – the idea of is that only the strongest firms survive. Richard Nelson and Sidney Winter came out from dissatisfaction of standard neo-classical economics to explain many empirical facts about long-term economic development and technological change. Evolutionary theory replaces profit-maximizing behaviour of firms with decision rules applied routinely over a period of time. Decision rules include routines for production, for managing workers, ordering inventory, advertising, or changing R&D. R&D in evolutionary theory has two fundamental mechanisms: search for better techniques and selection of firms by the market. R&D takes place if a firm is not satisfied with its current profits. R&D is more likely to yield results close to the current technology. Once a search is concluded, the firm decides whether or not the new technology is better. Concretely, profits are important even though profit maximization is not the explicit goal. The results of Evolutionary theory are similar to the Induced innovation theory. However, Evolutionary theory has one important difference, it is dynamic. Invention is cumulative. Successful R&D not only provides a new product, but also a new starting point for research. *De facto*, the exact nature of next year invention possibilities curve will depend from the outcome of this year research (R. NELSON, S. WINTER, *An evolutionary theory of economic change*, Belknap Press and Harvard Press, London and Massachusetts, 1982).

90 Firms balance the marginal costs and benefits of doing R&D. The marginal cost of R&D is the marginal cost of capital, i.e. the opportunity cost of using capital for R&D, rather than other investments. The marginal benefit is the marginal rate of return that is affected by the potential markets for the invention (demand-side influence) and the technological opportunities (supply-side influences). These affect the likelihood that a research project will be successful.

91 N. ROSENBERG, *Perspective on technology*, Cambridge University Press, 1976.

nologies (efficient lighting systems, renewable energy sources, collective urban transport, etc.). What are the most promising policies able to achieve the last step of the technological change process, i.e. to ensure a large deployment of new technologies? The IEA suggests that this deployment is coming from the market forces highlighted by the costs and risks associated to the diffusion of new technologies. Market barriers (in addition to technical barriers) impede a fast and large deployment of new energy technologies[92].

Within the induced technical progress, D. Finon insists on the competition among technologies, their selection and the competition among several futures.

> Les rendements croissants d'adoption contribue à accélérer les dynamiques de diffusion des futures techniques gagnantes en rendant plus rapidement obsolescentes les techniques en place, à faciliter la résistance de celles susceptibles de rebonds et à sortir du jeu les autres techniques candidates à l'entrée (...). La concurrence entre techniques et leur sélection qui s'opère sur ce mode dynamique engendrent un "multi-équilibre" ou, dit autrement, une concurrence de plusieurs futurs. La sélection entre techniques ou grappe d'innovation s'opère en dynamique soit sous l'impulsion de petits événements contingents en verrouillage progressif (...), soit par des politiques pour orienter l'adoption de nouvelles techniques et infléchir les infrastructures[93].

D. Finon also underlined that in most of the cases, Economics, Energy and Environment (E3) models are extrapolating historical tendencies. But these innovation dynamics are difficult to represent and to anticipate due to the induced technical progress.

Until the middle of the 1990s, E3 modelling has largely regarded technological change as an exogenous process, rather than an endogenous one[94]. As indicated by Alan Manne and Richard Richels, in Bottom-up models, technological progress is mainly an exogenous process related to cost and efficiency improvements of a relatively rich set of specific energy technologies. In Top-down models, economic output is given by a production function including technological progress, capital and labour, sometimes explicitly complemented by energy as production factors. Technology is often included in these macroeconomic models as a separate exogenously determined coefficient in

---

92 IEA, *Enhancing the market deployment of energy technology: a survey of eight technologies*, OECD, Paris, 1997.

93 D. FINON, *Prospective énergétique et modélisation – Identification de pistes de progression méthodologique*, *op. cit.*

94 The IIASA research team was among the first, in 1995, to include technological progress as an endogenous process in a Bottom-up model (MESSAGE).

the production function[95]. Applied to the climate change issue, Michael Grubb, Carlo Carraro and John Schellnhuber recognised this reality.

> In true Top-down models, supply side technologies were reflected in assumptions about the elasticity of substitution between generic carbon and non-carbon sources (if any), whilst an "autonomous energy efficiency improvement" (AEEI) parameter was used to reflect an assumed degree of decoupling between GDP and energy consumption – a single, fixed parameter encompassing both structural change in the relationship between economy and energy and the development and diffusion of demand-side technologies[96].

Jean-Charles Hourcade, Marc Jaccard, Chris Bataille and Frédéric Ghersi added that:

> if the input substitution elasticities critical to technological response in Top-down models are estimated from historical data, there is no guarantee that the values for these parameters would remain valid in a future with ambitious policies for environmental improvement, i.e. shaped by induced technological change. For example, until recently, there was no incentive to innovate and commercialize technologies with low greenhouse gas emissions. Today, such technologies are under development worldwide, providing households and firms with new choices that may change elasticities at the level of inter-factor substitution and the level of structural change, with significant implications for total economic output. Furthermore, while energy production and use can be treated as a marginal component of the overall economy in the short-run, in the long-run large changes in energy supply and use can have significant macroeconomic growth and structure implications. At the extreme; for very long term scenarios and in case of large departures from baseline projections, Top-down models cannot guarantee that their economic projections are underpinned by a feasible technical system[97].

D. Finon shows the possible progress in the representation of induced technological progress in Bottom-up and in Top-down models: in the first ones, by the introduction of the learning curves and in the second ones by the reduction of prices by public R&D, spillover rates from R&D and economies of learning:

> Les modèles Bottom-up représentent le progrès technique induit par l'introduction de fonctions de baisses de coût en fonction de la capacité installée de la technique innovante ainsi que du montant des investissements de

---

95 A. MANNE, R. RICHELS, *Environmental modelling and assessment*, Springer, The Netherlands, 1997.

96 M. GRUBB, C. CARRARO, J. SCHELLNHUBER, *Technological change for atmospheric stabilization: Introductory overview to the innovation modeling comparison project*, The Energy Journal, Special issue, IAEE, Cleveland, 2006.

97 J.C. HOURCADE, M. JACCARD, C. BATAILLE, F. GHERSI, *Hybrid modelling: New answers to old challenges*, *op. cit.*

R&D (...). Dans les modèles Top-down, en s'inspirant de l'esprit de la théorie de la croissance endogène, les perfectionnements visent à prendre en compte le progrès technique induit par le changement de prix à travers les investissements publics de R&D et les apprentissages cumulatifs. On introduit ainsi des coefficients de baisse de coût relatif du capital technique en fonction du stock de capital technique développé[98].

Since then, learning curves have been applied to several sectors, including energy technologies[99]. The International Energy Agency and the European Commission have put efforts on this issue[100]. EXTOOL and PHOTEX, two projects funded by the Fifth RTD Framework Programme dealt with experience curves of wind turbines and Photovoltaics[101]. In the European Commission WETO report using the POLES world energy model, the learning by doing effect and the impact of R&D on technology development have been considered. Learning curves have been established for power generation technologies (see Figure IV.6 where symbols indicate time steps of five years)[102].

98 D. FINON, *Prospective énergétique et modélisation – Identification de pistes de progression méthodologique*, *op. cit.*

99 A. LOSCHEL, *Technological change in economic models of environmental policy: A survey*, ZEW, Manheim, 2002.

100 International Energy Agency (C.O. WENE), *Experience curves for energy technology policy*, OECD, Paris, 2000; IEA, *Experience curves: A tool for energy policy analysis and design*, Workshop, Paris, 22-24 January 2003.

101 The EXTOOL consortium was composed of the Lund University (Lena Neij and Peter Helby), RISOE (Per Dannemand Andersen, Poul Erik Morthorst and Helge V. Larsen) and ISET (Michael Durstewitz). The PHOTEX project was coordinated by ECN (Gerrit Jan Schaeffer and Ad Seebregts), GENEC-CEA (Marion Perrin and Pascal Boulanger), Fraunhofer (Herman Laukamp), University of Utrecht-STS (Erik Alsema and Wilfried van Sark), and CESI (Carlo Zuccaro).

102 European Commission, World energy, technology and climate policy outlook, EC, Luxembourg, 2003. The WETO report was elaborated by ENERDATA (Bruno Lapillonne and Bertrand Chateau) and CNRS-LEPII (Patrick Criqui and Silvana Mima) in France, Bureau Fédéral du Plan (Dominique Gusbin) in Belgium, JRC-IPTS (Antonio Soria and Peter Russ).

Figure IV.6: Learning curves for power generation technologies up to 2030

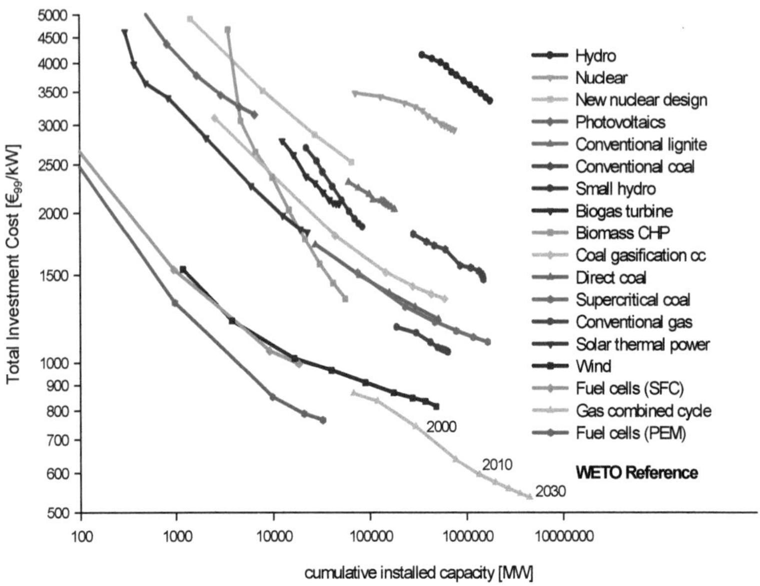

Both Top-down and Bottom-up models are used to assess the potential economic effects of energy-focused environmental policies, and particularly since the beginning of the 1990s, policies for mitigation of climate change. Because of their different structures and definitions of costs, the two types of models tend to predict different economic outcomes. Top-down models usually predict high – overestimated – costs of greenhouse gas emission reduction policies (constant elasticities of substitution and fixed autonomous energy efficiency index) while Bottom-up models usually predict low – underestimated – costs ("energy efficiency gap" and many cost-effective opportunities for emissions reduction)[103].

From a study performed by James Edmonds, it appears that from the perspective of long-term energy system development, the results from the Top-down and Bottom-up models are more closely linked to differences in assumptions regarding the economic cost associated with advanced technologies than to differences in modelling approach[104].

---

[103] M. GRUBB, J.A. EDMONDS, P. TEN BRINK, M. MORRISON, *The cost of limiting fossil fuel CO_2 emissions: A survey and analysis, op. cit.*

[104] EDMONDS, J.A., WISE M.A., MacCRACKEN C.N., *Advanced energy technologies and climate change: An analysis using the global change assessment model (GCAM)*, Pacific Northwest Laboratory, Richland, Washington, 1994.

The ACROPOLIS research project arrived at a similar conclusion[105]. Supported by the EU under the fifth RTD Framework Programme (See Chapter II), ACROPOLIS used up to fifteen world, European or national Economics, Energy and Environment (E3) models to simulate and evaluate selected policy measures and instruments and then to compare their impacts on energy systems essentially in terms of costs of greenhouse gas emissions (GHG) reduction and energy technology choice[106]. Four case studies were formulated considering policies and measures on renewable portfolio schemes and internationally tradable green certificates, emissions trading and global GHG abatement target, energy efficiency standards and internalisation of external costs. From a large set of quantified results, ACROPOLIS provided an international scientific consensus on some key issues[107]. The exercises from the three Top-down models (AIM, GEM-E3 and NEWAGE) and the three Bottom-up models (GMM, MESSAGE, POLES) concluded that the Kyoto targets could be achieved at a cost around 1% of GDP through global emissions trading, indicating also that this flexibility mechanism is a more cost-effective instrument for GHG mitigation than meeting the goal domestically without trade. In addition, it is worthwhile mentioning that the "Baseline" from the different world models were very similar up to 2020 in terms of CO_2 emissions (between 35,000 and 40,000 Million

---

105 DAS A., ROSSSETTI di VALDALBERO D., VIRDIS M., ACROPOLIS, *An example of international collaboration in the field of energy modelling to support greenhouse gases mitigation policies,* Energy Policy, Vol. 35, No. 2, February 2007. The ACROPOLIS (and later CASCADE-MINTS) initiative played an important role in peer-reviewing modelling functioning, data and assumptions in Europe – like the EMF (Energy Modeling Forum) in the USA.

106 The following models were used: AIM, DNE21, GEM-E3, GMM, MESSAGE, NEWAGE, POLES, MARKAL (MATTER, Nordic, MACRO-IT, Canada), PRIMES, NEMS, TIMES and NEO-MS.

Similar comparative studies are held by the International Energy Workshop (IEW) created in 1981 by Alan Manne and by the Energy Modeling Forum (EMF) hosted at the Stanford University Campus (John Weyant). Focused on the endogenous technical change in climate-economic models, the Innovation Modeling Comparison Project (IMCP) managed by Michael Grubb has also to be mentioned. At the EU-level, the follow-up of ACROPOLIS is the CASCADE-MINTS project funded under the sixth RTD Framework Programme coordinated by NTUA (Pantelis Capros and Nikos Kouvaritakis) and ECN (Martine Van Uyterlynde).

107 The European organisations participating to the ACROPOLIS project were IER in Germany (Alfred Voss and Anjana Das), ECN in The Netherlands (Remko Ybema and Koen Smekens), ENEA in Italy (Giancarlo Tosato), IIASA in Austria (Léo Schrattenholzer and Ger Klaassen), NTUA in Greece (Pantelis Capros and Nikos Kouvaritakis), PSI in Switzerland (Sokrates Kypreos), Chalmers University of Technology (Bosse Ryden) and the EU JRC-IPTS (Antonio Soria). NIES (Mikkio Kainuma) and RITE (Toshimassa Tomoda) from Japan, EIA from USA (Andy Kydes), GERAD from Canada (Richard Loulou) and IEA (Carmen Difiglio and Maria Virdis) also collaborated to the ACROPOLIS project.

tons of CO_2) and energy intensity (around 15 Giga Joule per € 1000 of GDP).

These issues have been subject of European energy socio-economic research (cf. Chapter II) concerning the dynamics of technical change. The European projects SAPIENT/SAPIENTIA and NEMESIS-ETC aimed to look at the role of R&D in inducing and accelerating technical progress in view of achieving certain policy objectives (non carbon emitting production growth, security of supply, etc.)[108]. These analyses incorporate learning processes and technological spillovers. A large part of the work consisted in the construction of an information base for energy technologies and the collection of techno-economic and R&D expenditure data for the estimation of the technology dynamics module. This module relates technological improvement to R&D and technology uptake. Concretely, technological improvements no longer fall as "manna from heaven" but depend on up-front investments.

108 The teams participating in these projects were NTUA in Greece (Pantelis Capros and Nikos Kouvaritakis), KUL (Stef Proost and Denise van Regemorter) in Belgium, CNRS/IEPE (Patrick Criqui and Silvana Mima), ECN (Koen Smekens) in The Netherlands, IER (Alfred Voss and Anjana Das) in Germany, IIASA (Leo Schrattenholzer) in Austria and PSI (Sokrates Kypreos) in Switzerland; Centrale Recherche and CCIP (Paul Zagamé, Arnaud Fougeyrollas and Pierre Le Mouël) in France, University of Amsterdam (Reyer Gerlagh), Universität Hamburg (Richard Toll), CESI and FEEM (Marzio Galeotti and Valentina Bosetti) in Italy.

CHAPTER V

European Quantitative Tools

Summary

The objective of this chapter is to present the different specificities and mechanisms of the major tools developed by European economic energy research. These tools (GEM-E3, POLES, PRIMES, SAFIRE, MURE and EXTERNE) will be the most often quoted scientific reference in the EU energy-related policies between 1995 and 2005 (see Chapter VI).

As presented in Chapter II (*Energy research in the EU*), the complex interactions between Economics, Energy and Environment (E3) have been addressed from a long-time in a systematic way at the European Union level. From the 1980s, first generation models like MEDEE, EFOM, MIDAS or HERMES have been supported by European research activities. At the very beginning of the 1990s, European research work started – in cooperation with the USA – on the accounting framework on fuel cycles (quantification of various externals costs associated with the production and consumption of energy from different fuel sources). At this moment, European research activities also launched database-related activities for energy efficiency measures and for renewable energy sources.

In order to better investigate the impacts of different E3 policies and measures (taxation, renewable, energy efficiency, greenhouse gas emission trading, etc.), "new generation models", macro-economic models, externalities accounting framework and databases have been developed in the 1990s.

The years 1991-1992 will see the emergence of the first version of the PRIMES and POLES partial equilibrium models for the European and world energy systems. In the same time, the general equilibrium model for energy, economy and environment GEM-E3 will be built. In that period, the EXTERNE accounting framework will start to be implemented to concrete case studies (power plants in some countries). The MURE and SAFIRE European databases and softwares for energy efficiency measures and for renewable energy sources will also be developed at this period[1].

In the middle of the 1990s, two major European research projects (See Chapter II) had a major role in "clustering" modelling activities, in achieving a modelling scientific consensus and in giving a political visibility to these E3 and external costs issues. The two projects were the "Climate technology strategy within competitive energy markets: towards new and sustainable growth" and the "National implementation in the EU of the EXTERNE accounting framework".

This chapter presents the main models and tools developed at the EU-level (GEM-E3, POLES, PRIMES, SAFIRE, MURE and EXTERNE), with their main characteristics and their basic scheme. As a conclusion, these models and tools have in common that they are all quantitative, that they have been developed at the EU-level under the "Economic research in energy" activity of the European Community Research Framework Programmes (see Chapter II), that they cover the whole EU and that they tackle the interactions among energy, economy and environment. These tools will provide key-inputs to some major EU energy-related policies (see Chapter VI).

European macro-economic model GEM-E3

GEM-E3 (*General Equilibrium Model for Energy, Economy and Environment*)[2] is a large-scale computable general equilibrium model simultaneously representing the EU15 Member States (without Luxembourg) that provides details on the macro-economy and its interaction with the energy and environment system. GEM-E3 computes the equilibrium prices of goods, services, labour and capital according to the Walras law[3]; it uses the software PATH in GAMS.

---

1 European Commission, *JOULE I – catalogue of contracts*, *op. cit.*; European Commission, *JOULE II – project synopses*, *op. cit.*; European Commission, *JOULE III – project synopses*, *op. cit.*; European Commission, *Socio-economic projects in energy and environment*, *op. cit.*; European Commission, *Energy – Economic, social and policy-oriented research in Europe*, *op. cit.*

2 European Commission, *GEM-E3 computable general equilibrium model for studying economy-energy-environment interactions*, OPEC, Luxembourg, 1995.

C. BOEHRINGER *et al.*, *European emission mitigation policy and technological evolution: Economic evaluation with the GEM-E3*, 2000, Final Report to the European Commission (DG Research).

3 M.E.L. WALRAS, *Éléments d'économie politique pure* (*Elements of Pure Economics*), 1874. Walras provided one of the first comprehensive mathematical analyses of general economic equilibrium. For each product there is a "demand function" that expresses the quantities of the product that consumers demand as depending on its price, the prices of other related commodities, the consumers' incomes, and their tastes. For each product there is also a "supply function" that expresses the quantities

Concretely, GEM-E3 includes all simultaneously interrelated markets and represents the system at the appropriate level with respect to geography, the sub-system (energy, environment, economy) and the dynamic mechanisms of agent's behaviour. It formulates separately the supply and demand behaviour of the economic agents, which are considered to optimise individually their objective while market derived prices guarantee global equilibrium. Prices are computed by the model because of supply and demand interactions in the markets. A version of GEM-E3 (*GEM-E3-World*) has also been developed for covering the world (divided in 21 regions).

The model is simultaneously multinational (for the EU) and specific for each country which allows European or country-specific policy analysis. Although global, the model exhibits a high degree of disaggregation concerning sectors, economic agents, structural features of energy/environment and policy-oriented instruments (e.g. taxation). It allows the consistent evaluation of distributional effects of policies. It is a multi-period model, involving dynamics of capital accumulation and technological progress, stock and flow relationships and backward looking expectations. In addition, the model covers the major aspects of public finance including all substantial taxes, social policy subsidies, public expenditures and deficit financing, as well as policy instruments specific for the environment/energy system. The model determines the optimum balance of energy demand and supply, atmospheric emissions and pollutant abatement, simultaneously with the optimising behaviour of agents and the fulfilment of the overall equilibrium conditions. In this sense, the model analyses the interactions between the economy, the energy and the environment systems.

The results of GEM-E3 include projections of full input-output tables by country, national accounts, employment, and capital flows, balance of payments, public finance and revenues, household consumption, energy use and supply, and atmospheric emissions. The computation of

---

producers will supply as depending on their costs of production, the prices of productive services, and the level of technical knowledge. In the market for each product there is a point of "equilibrium" at which a single price will satisfy both consumers and producers. Since equilibrium in each market depends on what happens in all other markets, finding a "general equilibrium" solution involves the simultaneous determination of equilibria in all markets. For this purpose, Walras constructed a mathematical system of simultaneous equations reflecting demand and supply in all the markets. Walras supposed a perfect competition: large number of producers and consumers on the market (no one is sufficiently strong to influence the results on the market); free entrance (no barriers for new agents to enter or to leave the market); homogeneity in the products (competition can therefore only be done through price); free movement of labour and capital (from one company to another or from one market to another); perfect information (everybody knows in the same time the quantities offered and demanded by all the agents at a certain price).

equilibrium is simultaneous for all regions covered by the model and foreign trade links.

GEM-E3 provides a consistent evaluation of distributional effects across countries, economic sectors and agents. The burden sharing aspects of energy supply and environmental protection are fully analysed, while ensuring that the economy remains at a general equilibrium condition.

GEM-E3-Europe includes 18 products and sectors: Agriculture, Coal, Oil, Natural Gas, Electricity, Ferrous & Non-Ferrous Metals, Chemical Products, Other Energy-Intensive Industries, Electrical Goods, Transport Equipment, Other Equipment Goods Industries, Consumer Goods Industries, Building and Construction, Telecommunication Services, Transport, Services of Credit and Insurance, Other Market Services and Non-Market Services.

GEM-E3 represents:

- Four economic agents – households, firms, government and foreign sector;
- Two primary production factors – labour and capital;
- Nine government revenue and income flow categories – direct taxation, indirect taxation, energy and environmental taxation, *ad valorem* taxation, property taxes, capital taxes, social security, social benefits, subsidies (production and exports), import duties and foreign sector transfers, and social security and revenues from government enterprises;
- Thirteen household expenditure categories – Food, Beverages and Tobacco, Clothing and Footwear, Housing and Water, Fuels and Power, Housing Furniture and Operation, Heating and Cooking Appliances, Medical Care and Health Expenses, Transport Equipment, Operation of Transport Equipment, Purchased Transport, Telecommunication services, Recreation, Entertainment, Culture, etc. and other services;
- Three pollutant emissions for Europe (CO_2, SO_2, NO_x) and greenhouse gases in GEM-E3-World.

The most important results provided by GEM-E3 are:

- The dynamic annual projections in volume, value and deflators of national accounts by country;
- The Input-Output tables by country and for EU as a whole, for the sectors mentioned above;
- The distribution of income and transfers in the form of a social accounting matrix by country; the employment, capital, and investment by country and sector;

- The atmospheric emissions, pollution abatement capital, purchase of pollution permits and damages;
- The consumption matrix by product and investment matrix by ownership branch;
- Public finance, tax incidence and revenues by country/region; and
- Full trade matrix for EU and the rest of the World (GEM-E3-Europe).

Figure V.1: Basic scheme of the GEM-E3 model

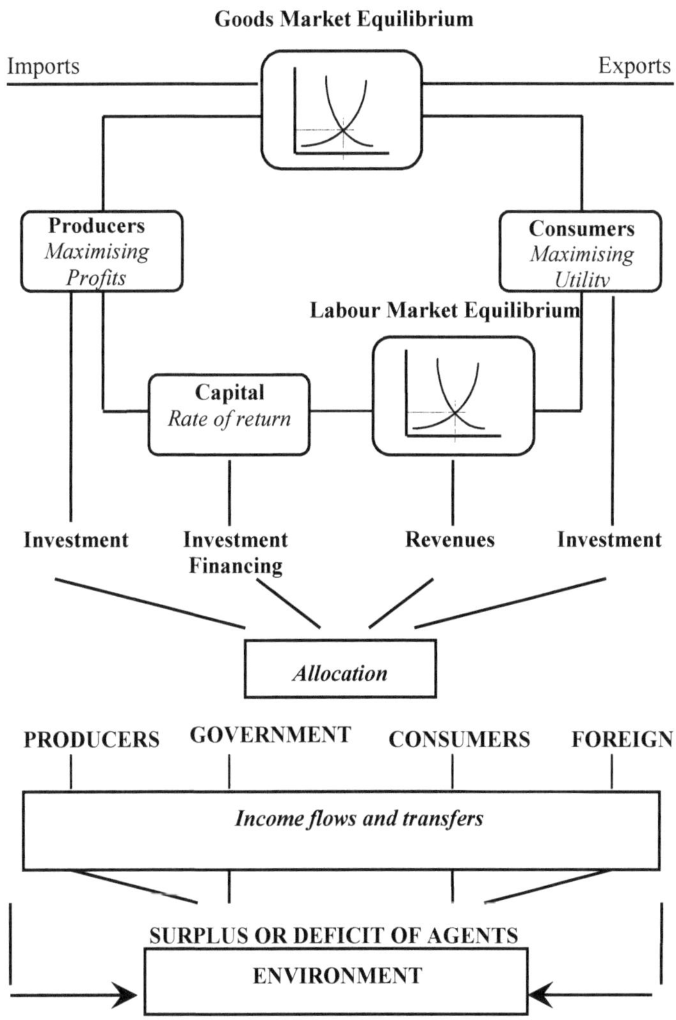

World energy model POLES

POLES (*Prospective Outlook on Long-term Energy Systems*) is a partial equilibrium model of the world energy system to 2050. It has been designed in the framework of a hierarchical structure of interconnected sub-models at the international, regional and national level[4]. The dynamics of the model is based on a recursive (year by year) simulation process of energy demand and supply with lagged adjustments to prices and a feedback loop through international energy prices.

The POLES model provides information on energy flows for each country/region in a structure similar to that of a standard IEA-type energy balance. A summary balance provides synthesised information on energy consumption and transformation, new energy technologies and electricity production capacities.

In the current geographic disaggregation of the POLES model, the world is divided into 47 countries/regions. The largest countries in terms of energy and emissions are identified and treated with a detailed model. In the current version, these countries are the G8 countries and all EU Member States plus five key developing countries: Mexico, Brazil, India, South Korea and China. The remaining sub-regions are dealt with more compact and homogeneous models.

For each region, the model articulates four main modules dealing with: final energy demand by main sectors; new and renewable energy technologies; the conventional energy and electricity transformation system; and the fossil fuel supply.

In the detailed demand models, the consumption of energy is disaggregated into 15 homogeneous sectors that allow the identification of the key energy-intensive industries, the main modes of transport and the residential and tertiary activities.

Twelve technologies are considered in the "new and renewable energy module", from biomass gasification to fuel cells, wind and photovoltaics. Twelve technologies are also considered for large-scale power generation form "clean coal" to GTCC (Gas Turbine Combined Cycle) and "Third generation" nuclear power plants.

The model has a comprehensive and detailed structure – both in terms of regional and sectoral disaggregation. It allows determining the fundamental variables in energy market developments and in inter-technology and inter-fuel competition. The predominance of price

---

4 European Commission, *POLES*, OPEC, Luxembourg, 1996. CRIQUI P., MIMA S. and VIGUIER, *Marginal abatement costs of CO_2 emission reductions, geographical flexibility and concrete ceilings: an assessment using the POLES model*, Energy Policy, Vol. 27, 1999.

adjustment mechanisms in every module of POLES allows consistent simulations of environmental policies to be produced, through the introduction of "penalties" for greenhouse gas emissions. These penalties can be the result either of environmental taxes or of emission permit systems.

The model has been initially designed in order to provide simulations of the impacts of long term energy technology scenarios, but the developments rapidly aimed at providing detailed economic assessments of abatement policies and of their consequences on energy technologies' performances and markets. For example, the model produced marginal abatement cost curves (at national or sectoral level), in order to study the impacts of flexibility mechanisms and the structure of tradable emission permit markets.

Being a partial equilibrium model with exogenous world economic projections, the POLES model, however, does not account for macro-economic feedbacks and indirect impacts of energy and environment policies (cf. GEM-E3 model below).

Figure V.2: Basic scheme of the POLES model

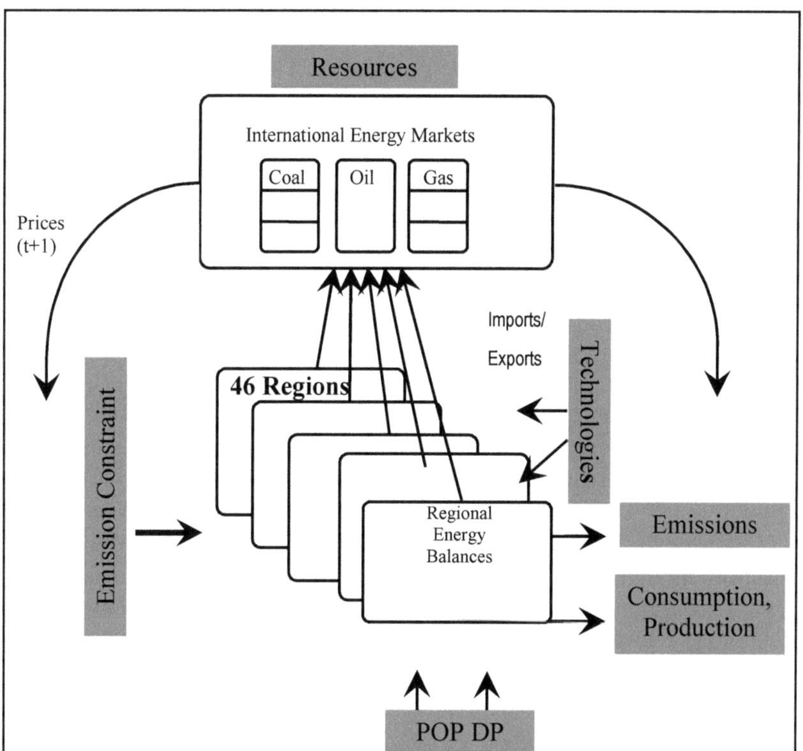

The energy balance data of the POLES model are extracted from the ENERDATA database that stems from national and international statistical sources (UN, OECD-IEA, Eurostat). International macro-economic data concerning GDP, the structure of economic activity, deflators and exchange rates come from the CEPII-CHELEM database that is based on the United Nations, World Bank and International Monetary Fund data. Techno-economic data (energy prices, equipment rates, power plant capacities, costs of energy technologies, etc.) are gathered both from international and national statistics.

Most development and uses of the POLES model have been performed with the TROLL software, while a VENSIM version of the model has been recently developed.

European energy model PRIMES

PRIMES (*Price Induced Model of the Energy System*) is a partial equilibrium model for the European Union energy systems[5]. It simulates a market equilibrium solution for energy supply and demand in the 27 Member States of the European Union. The model determines the equilibrium by finding the prices of each energy form such that the quantity supplied matches the quantity demanded (both producers and consumers respond to changes in prices). The equilibrium is static (within each time period) but repeated in a time-forward path, under dynamic relationships. PRIMES is used for forecasting, scenario construction until 2030 (in 5-year intervals).and policy impact analysis.

Although behavioural and price driven, PRIMES simulates in detail the technology choice in energy demand and energy production. The model explicitly considers the existing stock of equipment, its normal decommissioning and the possibility for premature replacement. At any given time, the consumer or producer selects the technology of the energy equipment on an economic basis that can be influenced by policy, market conditions and technology changes. The model is technology-rich and is suitable for exploring the possibilities offered by technological developments including the impact of policies that induce technical change. Due to its high level of disaggregation, it is capable of simulating permit market configurations to a high degree of refinement.

Environmental impacts are also integrated into PRIMES. The optimisation modules simultaneously consider energy and environmental costs. Constraints are built in to represent environmental regulation. The technology choice mechanisms also consider abatement equipment. The

5 European Commission, *PRIMES*, OPEC, Luxembourg, 1995; European Commission, *Energy and transport trends to 2030*, *op. cit.*; P. CAPROS *et al.*, *Climate technology strategy 1* – Controlling greenhouse gases, policy and technology options, *op. cit.*

main policy instruments for the environment, as considered in PRIMES, are regulations by sector and by country, environmental taxation, pollution permit market and subsidisation of abatement costs for electricity and steam.

The demand is represented by four sectors:

- Residential (different categories of dwelling defined according to the main technology used for space heating; dwellings are further subdivided in energy uses in which the electric appliances for non-heating and cooling are considered as a special sub-sector);
- Commercial and agriculture (more than 30 end-use technology types);
- Industry (iron and steel, non-ferrous metals, chemicals, building materials, paper and pulp, food drink tobacco, engineering, textiles and other industries); and
- Transports (passenger and goods transport as separate sectors which are further subdivided in sub-sectors according to the mode of transport and with alternative technologies for each mode).

The energy supply is represented by three modules: Electricity production (with hundreds of power plant types – thermal, reservoir and intermittent, with chronological load curves, interconnections, network representation, cogeneration of power and steam and district heating); Refineries (cracking, reforming etc.); and Natural gas (regional supply detail for Europe, Russia, Middle Africa, North Sea, etc.; transportation and distribution network).

PRIMES includes 24 energy fuels in total: Coal, Lignite and Peat, Crude Oil, Residual Fuel Oil, Diesel Oil, LPG, Kerosene, Gasoline, Naphtha, Other oil products, Bio-fuels, Natural and derived gas, Thermal Solar (active), Geothermal, Steam (industrial and distributed heat), Electricity, Biomass and Waste, Hydrogen, Solar electricity, Wind, Hydro.

Until 2030, PRIMES provides energy balance sheets per country and per year; energy costs, producer and consumer prices; power generation park, load curves, load factors, investment and marginal costs (central systems, combined heat-power, exchanges); refining units, expansion, costs; natural gas transport and distribution: flows, capacities, costs; endogenous treatment of energy savings and new technologies; atmospheric emissions (CO_2, NO_x, SO_2), abatement equipment and standards.

PRIMES uses the software GAMS v. 2.25 with PATH solver and Cplex, MS EXCEL.

Figure V.3: Basic scheme of the PRIMES model

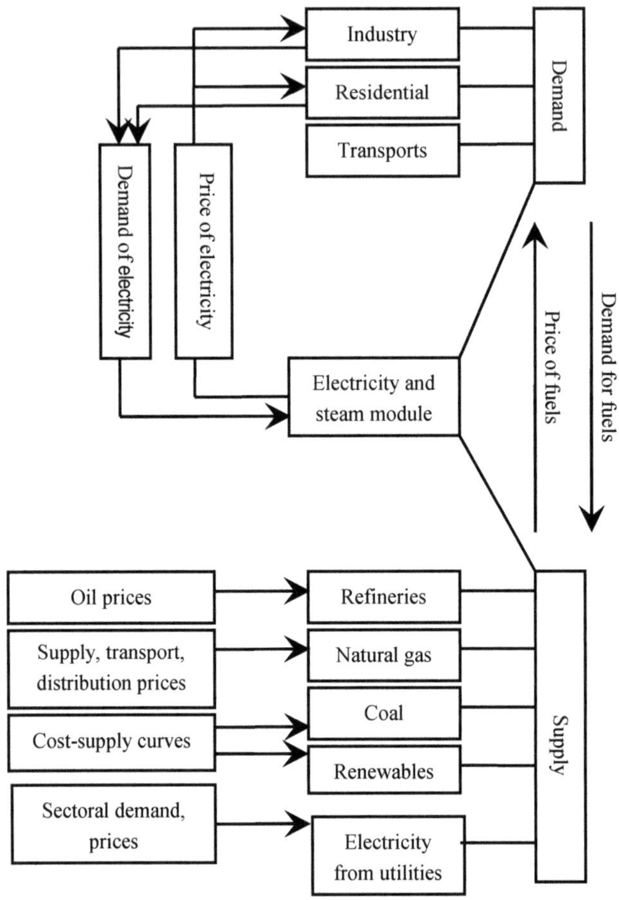

European software for renewable energy SAFIRE

SAFIRE (*Strategic Assessment Framework for the Implementation of Rational Energy*) is an engineering-economic Bottom-up supply and demand model for the assessment of first-order impacts of rational (i.e. renewable and new non-renewable) energy technologies on a national, regional or local level against a background of different policy instruments and scenario assumptions[6].

[6] European Commission, *SAFIRE*, OPEC, Luxembourg, 1995. WHITELEY M., M. BESS, *Renewable energy technologies and the European industry*, International Journal of Global Energy Issues, Vol. 14, 2000.

The range of countries within SAFIRE include the EU-25, four other Western European countries (Cyprus, Iceland, Norway, Switzerland) and eight other countries worldwide (Brazil, Canada, China, India, Indonesia, Japan, Mexico, USA). Local municipalities are also included for Austria, Bulgaria, Denmark, Germany, Ireland, Netherlands, Poland, Portugal and United Kingdom.

SAFIRE is data-driven and the database is divided into two areas, which is the base year data (based upon actual statistics) and the scenario data (a view of the future). For a given location and scenario, the primary SAFIRE calculation uses a substitution methodology to assess the potential future supply of rational energy, whilst matching a demand matrix defined by the user. If the demand-side management module is activated, this is calculated first, as it directly affects domestic sector demand. In the main SAFIRE calculation, priority is given to *decentralised* heat and electricity generation and, in a second stage, to *centralised* heat and electricity generation (district heating and electricity). Decentralised is defined as the primary output being generated for on-site consumption.

On the supply side, the software includes an extensive database for 22 renewable energy technologies (RETs), eight new non-RETs and seven fuelling options for co-generation plants including fuel cells. All of these technologies, where relevant, can be applied as electricity only plant, thermal plant or as cogeneration. In addition to these technological options, they can be applied as decentralised or centralised (both electricity and district heating) generation. SAFIRE also includes two transport biofuels (bio-ethanol and biodiesel).

On the demand side, SAFIRE represents the: Residential sector (a single average category of dwelling in which energy consumption is divided into space heating (thermal/electric), water heating (thermal/electric) and electricity; the demand side management module further divides electricity consumption into cooling, lighting, insulation and other electrical appliances); Commercial sector (small, medium, large and very large, based upon floor area with specific energy consumptions for electricity, space heating and water heating); Industry: (divided into 43 predefined NACE categories with specific energy consumptions for process heating (both high and low temperature), water heating, space heating, electricity and cooling); Agriculture (with specific energy consumptions for process heating, low temperature heating, space heating, electricity and cooling); Transport (with calculation of the potential for bio-ethanol and biodiesel as substitutes for petrol and diesel).

SAFIRE includes 10 biomass/waste fuels (forest residues, energy crops – wood and biofuels, solid agricultural wastes, liquid agricultural wastes, municipal solid wastes, sewage gas, landfill gas, solid industrial wastes, liquid industrial wastes) and 11 conventional fuels (Electricity – peak and off-peak, natural gas, derived gas, coal, lignite/peat, heavy fuel oil, light fuel oil, petrol, diesel, and nuclear).

The SAFIRE calculation is divided into eight stages: Energy demands; Demand side management calculation; Renewable energy technical potential; Renewable and non-renewable energy market potential (decentralised only); District heating market potential and penetrations; Renewable and non-centralised non-renewable market penetration; Centralised electricity market penetration; and Cost benefit analysis.

The fuelling options open to co-generation, which is assumed to be all decentralised (district heating co-generation is treated separately). The most important feature of the co-generation methodology is the calculation status of liberalisation on the electricity market. This directly affects the revenues received for co-generation output and hence the size of plant constructed.

SAFIRE can be applied to assess the impact of energy technology and associated policies on a number of economic indicators like market penetration; net employment creation; pollutant emissions (CO_2, SO_2, NO_X, VOC and PM); import dependency; capital expenditure; external costs and government expenditure. These indicators are calculated as *net* effects, taking into account the impact of the new technology on the one hand and the impact caused by the displaced conventional technology on the other.

The main outputs of SAFIRE (Dynamic Annual Projections in Specific Units) are: Energy demand (heat, electricity, transport) by sector (GWh); Technical potential – by renewable technology (GWh); Decentralised renewable market potential – by technology, sector, location, heat/electricity – (GWh, Mtoe); Market penetration – by technology, sector, location, heat/electricity, centralised/decentralised (GWh, Mtoe); Net Employment (by technology, sector and location in net jobs/year); Net emissions – CO_2, NO_x, SO_2, CO, VOCs, particulates (by technology, sector and location in net change in tonnes of emission/year); Import dependency (by technology, sector and location in GWh/year, Euro/year); Capital investment (by technology, sector and location in Euro/year); External costs (by technology, sector and location in Euro/year); and Government expenditure (by technology, sector, location in Euro/year).

SAFIRE provides through its data-bases a systematic procedure for compiling cost and performance data for new technologies. This is an important consideration for long-term modelling studies because different assumptions regarding the penetration of new technology can lead to major divergences in results when extrapolated over a long-term. It is important to be able to distinguish the consequences of different behavioural assumptions and different numerical data.

SAFIRE runs (under FoxPro 2.5b software) for a 40 year period from a flexible starting year specified by the user. Outputs are saved at 5-year intervals.

Figure V.4: Basic scheme of the SAFIRE renewable software

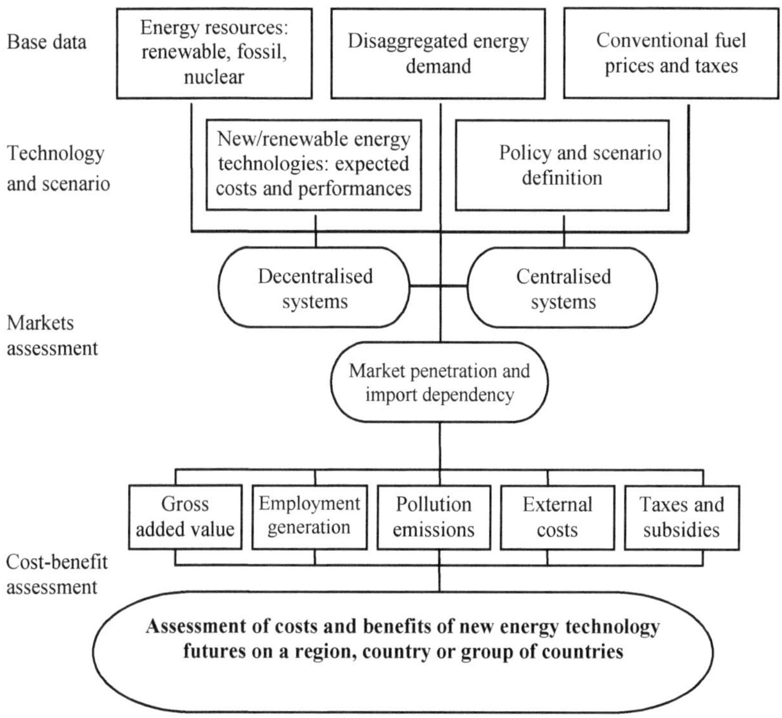

European software for rational use of energy MURE

MURE (*Mesures d'Utilisation Rationnelle de l'Energie*) is a Bottom-up, technology-related simulation tool which contains a database of energy saving and environmental measures in the EU Member States (and Norway) and a database of energy-related statistics which describes the energy situation of each country[7].

---

[7] Fraunhofer-ISI, *A comparison of thermal building regulations in the European Union*, MURE database case study No. 1, EC SAVE programme, 1999.

MURE is particularly aimed at providing estimates of the economic energy saving potentials resulting from the implementation of national or European energy policies. It covers the following four sectors: Household; Transport; Tertiary and Industry.

MURE features three main components:

Firstly, an energy saving intervention quantitative database in which all the relevant measures enacting the energy policy of each Member State are described in terms of measure implementation dates, estimates of the involved stock and of the penetration rate, the number and type of energy saving interventions that the measure implies (e.g. insulation, maintenance, boiler substitution, new technologies, etc.) and the interventions absolute gain (rate of energy saving).

Secondly, a quantitative database of energy related statistics, technical parameters and investment costs covering the EU Member States and broken down by end-use sector. For household for example, this concerns heating, sanitary water and electric uses.

Thirdly, the simulation tool carries out calculations of energy savings and emissions reduction potentials.

All data refer to a reference year. The setting of a simulation exercise requires the specification of the duration and of the number of the simulation steps, for example, a "back-casting" exercise may refer to the year 1990 and have 5 simulation steps of two years each in order to arrive to the year 2000, while a forecasting exercise can start from the year 2000 and have 5 calculation steps of 5 years each in order to target the year 2025.

Once the measure quantitative description is set up, the tool automatically produces two simulation curves: the reference case, that takes into account all the measures implemented before the reference year and still in force, and the policy scenario, that considers all the measures implemented after the reference year. It is also possible to select the measures that are considered in the policy scenario in order to appreciate separately the impact of each of them.

For the reference year, for households, the following data are required: dwelling stock and dwelling unitary energy consumption by age and fuel; hot sanitary water devices stock and unitary energy consumption by fuel; appliances data (stock and energy consumption of the stock); unitary emission coefficients by energy source; fuel mix and total primary energy consumption of the national electricity production; primary electricity conversion coefficient (toe/kWh); fuel mix and total primary energy for the district heating heat production; energy efficiency of the district heating system.

The MURE time series include the annual number of household and the annual number of new individual and collective dwelling (from the date of the first building code measure).

For the reference forecast figures, MURE uses the total household electricity consumption growth rate (figures coming from the PRIMES model); the electricity production fuel mix, the district heating fuel mix and the household growth rate (figures coming from the POLES model).

The measures are analysed and parameterised by homogeneous category (fiscal, informative, etc.) and by issuing date. The simulation criteria are therefore set according to the type of intervention to be simulated (insulation, boiler substitution, etc.), the simulation of the measure relative gain (% saving) and the definition of the measures penetration rate (i.e. the rate of penetration in the involved dwelling stock).

The technical parameters of MURE are the stock average and new heating system efficiencies; the stock average insulation parameters; the equipment costs (boilers, insulation works, new equipments, appliances, etc.) and other data such as the equipment life span, the interest rates, etc.

The impact simulation methodology is done in four steps: general data set up; measures analysis and parametrisation; run and discussion of results; and possible further measures parametrisation and data calibration.

The main simulation criteria take into account the parameters in: the regulating laws for insulation; the renewal and substitution rate for boilers (18 years for boilers that serves more than three dwellings and 15 years for others boilers); the average gain percentage for generic intervention on the heating system (burner substitution, device controls, individual metering, etc.); and the penetration rates of the financial and other similar measures (accelerated by fiscal and tariffs measures).

The outputs of MURE (household) are provided in terms of global primary energy consumption per year (back casting or forecasting); of total primary energy consumption by fuel per year; of electricity production mix per year, of emissions per year and of economic assessment. Concretely, these outputs allow taking into account the impact of the measures that modify, at the final consumer level, the energy supply mix through e.g. increased recourse to micro CHP, fuel cells or PV installations.

Figure V.5: Basic scheme of the MURE (household) rational use of energy software

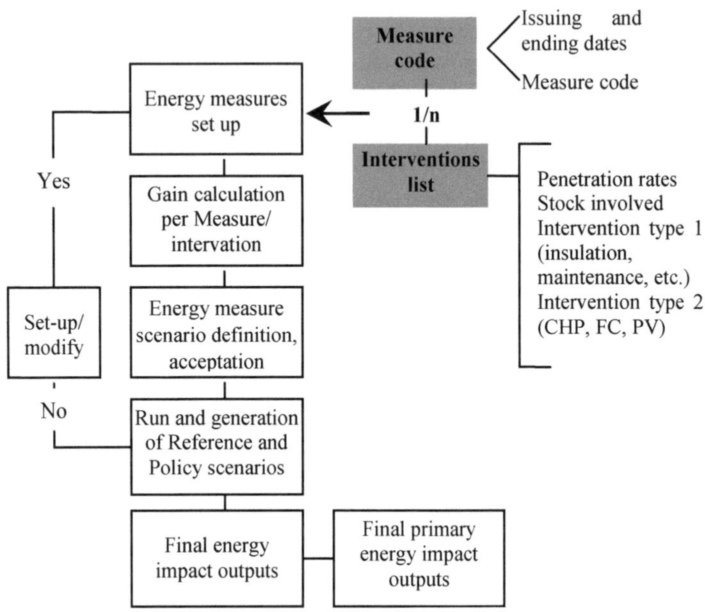

European external costs accounting framework EXTERNE

The EXTERNE (*Externalities of Energy*) methodology aims at assessing the impacts of energy and transport systems and technologies[8]. The impacts of operation and life cycle of the analysed technologies on human health, materials, plants and ecosystem are quantified and converted into monetary units by using contingent valuation, i.e. by assessing the willingness to pay of the population to avoid the quantified damages. As damages are site-specific, a Bottom-up methodology, the "impact pathway approach" has been developed.

The "impact pathway approach" starts with the emissions of pollutants and then calculates the changes in ambient concentrations and

[8] European Commission, *ExternE: Externalities of Energy – Summary* (Vol. 1), *Methodology* (Vol. 2), *Coal and lignite* (Vol. 3), *Oil and gas*, (Vol. 4), *Nuclear* (Vol. 5), *Wind and hydro* (Vol. 6), *Methodology 1998 update* (Vol. 7), *Global warming* (Vol. 8), *Fuel cycles for emerging and end-use technologies, transport and waste* (Vol. 9), *National implementation* (Vol. 10), OPEC, Luxembourg, 1995 and 1999. European Commission, *ExternE: Externalities of Energy – Methodology 2005 update*, OPEC, Luxembourg, 2005.

depositions caused by these emissions. Concretely, an atmospheric dispersion model is used to identify the source of pollution in terms of the site of pollution and the technology responsible and to determine the rate of emission, for example of particulates in kilogramme/year. Thus, it determines the increase in concentration at the receptor sites.

In a next step, exposure-response relationships are used to quantify the physical impacts on:

- human health (reduction in life expectancy due to long and short term exposure to pollutants; morbidity effects such as chronic bronchitis, non-fatal cancer, congestive heart failure, asthma attacks, respiratory hospital admissions, cough for different parts of the population, e.g. asthmatics, children, adults, population above 65 years, and for the total population);
- on building materials (ageing of galvanised steel, limestone, mortar, sandstone, paint, rendering, and zinc for utilitarian buildings);
- on crops (yield change for wheat, potato, sugar beet, rice, tobacco, sunflower seed, etc., increased need for liming and fertilising effects) and
- on ecosystems (exceedance of critical loads for acidification and eutrophication for terrestrial as well as acidification for aquatic ecosystems.).

In the last step of the "impact pathway approach", physical impacts are valued in monetary terms. This monetary valuation provides, mainly for power generation and the transport sector, a figure corresponding to the so-called "external costs" (in Euro cents per kilowatt-hours for electricity and, e.g., in Euro cents per vehicle – kilometre for transport). This figure has been calculated for case studies around all of the European Union. As damage occurs on the local, European and global level, all these scales are considered in the analysis.

The EXTERNE methodology can been applied to assess external costs of a large number of electricity fuels and technologies (coal, gas, nuclear and renewables) and transport technologies for all modes (rail, road, waterways, air) operated in various locations.

To cover pollutant dispersion and chemical transformation on different scales, the ECOSENSE software (MS-Windows based software system) is used. ECOSENSE is an integrated software tool for the assessment of airborne pollutants. Based on European data for receptor distribution (population, crops, building materials), background emissions (amount and spatial distribution), and meteorology (precipitation, wind speed and wind direction), ECOSENSE allows to calculate location specific marginal external costs of a stationary source (e.g. a power plant) due to emissions of air pollutants. The input required is annual emissions of NO_x, SO_2, PM_{10}, NMVOC, CO_2, N_2O, CH_4; the pollutants considered are O_3, SO_2, PM_{10}, sulphates, nitrates and greenhouse gases. All geographical information is organised using the EMEP co-ordinate system, which defines grid cells of 50x50 km^2 covering all European Union countries.

Impacts from greenhouse gas emissions causing global warming are taken into account using a damage cost as well as an avoidance cost approach; that means that marginal damage costs of emitting CO_2 and other greenhouse gases are estimated in addition to the marginal avoidance costs of reaching the Kyoto targets.

Impacts, where pollutants are transported through soil and water, before causing damage to human health, animals or plants, are analysed with a tool that simulates the path of the pollutants from the deposition of air pollutants to soil and water and from direct emissions to water and soil via ground water to agricultural plants, fish and drinking water and further to human beings.

The monetary values for assessing the different studies are derived from a thorough analysis of existing contingent valuation studies. For identifying results to monetise the "Value of a Statistical Life" (evaluated at € 1 Million) and the "Life of Years Lost" (evaluated at € 40,000), surveys of EU population have been conducted[9].

A detailed analysis of uncertainties in the applied impact pathway assessment has been carried out in the EXTERNE project series, including comparisons of modelled with measured concentration data and sensitivity analyses. With its user interfaces for changing all relevant input data, EcoSense supports sensitivity analyses by the user allowing for easy modification of assumptions. Uncertainties are still quite high.

---

9 D. ROSSETTI di VALDALBERO, J.M. BAER, *Les coûts sociaux en Europe – Recherche, énergie, transport et environnement*, IEPF, No. 74, Québec, 1er trimestre 2007.

Figure V.6: Basic scheme of the EXTERNE impact pathway approach

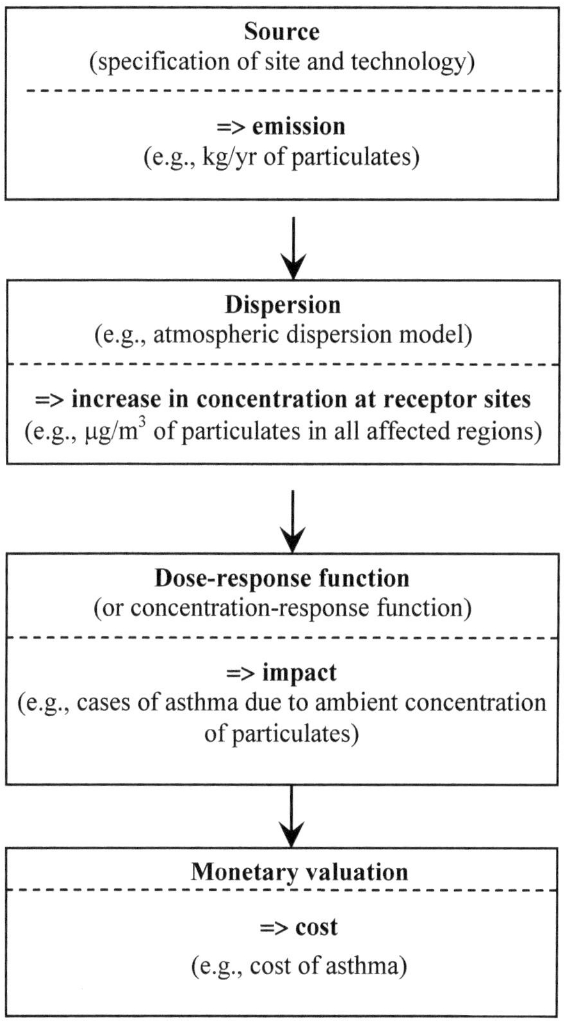

CHAPTER VI

Case Studies on EU Policies

Summary

The tools mentioned in this Chapter (GEM-E3, POLES, PRIMES, HERMES, MIDAS, E3ME, SAFIRE, MURE and EXTERNE) and used to support various EU energy-related policies have been originally elaborated by researchers under various EU RTD Framework Programmes (see Chapters II, III, IV and V).

This Chapter gives examples on how the results of these tools developed by European scientists, essentially in the economic discipline, have influenced policy decisions at the European level. In a certain way, this Chapter highlights the "origin of the figures" that will be adopted as a target or as a reference basis in the EU energy-related policies.

As empirical analysis, five European political case studies in the period 1995-2005 are addressed in this Chapter: the Community framework for energy taxation; the EU Climate Change policy; the promotion of renewable energy sources in Europe; the energy efficiency policies in the EU; and the measurement and internalisation of external costs in Europe[1].

In the Community framework for energy taxation, three models developed by European researchers under the second, third and fourth RTD Framework Programmes (HERMES, GEM-E3 and E3ME) have been used to assess the macro-economic impacts of the proposed Directive on the taxation of energy products in 1997. The three models concluded that when recycled to reduce the social security contributions the proposed increase in the EU minimum rate on motor fuels, heating fuels and electricity would have very marginal – and positive – impacts on GDP (from 0.06% to 0.2%) and employment (from 155,000 to 457,000 jobs).

For the revised 2003 proposal, only the GEM-E3 model was used. It concluded that approximately no change in GDP and employment

[1] Some insights have also been given to more recent EU policies. Details on the 2007 EU energy and climate change package can be found in: D. ROSSETTI di VALDALBERO, *Les nouveaux paradigmes énergétiques et environnementaux: la place de l'Europe*, Revue de l'Énergie, No. 584, juillet-août 2008.

would occur from the Community framework for energy taxation (respectively from -0.01% to 0.00% for GDP and from -0.01% to 0.02% for employment). If these results were not sufficient to argue in the sense of an ambitious minimum level of taxation (the final 2003 Council Directive was very modest in comparison to the 1997 Commission Proposal), these modelling studies helped in achieving an agreement at the Council.

In the EU Climate Change policy, the PRIMES model developed by European economists in the second, third and fourth RTD Framework Programmes provided some crucial data for the pre-Kyoto and post-Kyoto EU and international negotiations.

Firstly, PRIMES gave the main reference for the future increase in CO_2 emissions. The "Baseline" expected an increase of 8% of CO_2 emissions by 2010 in relation to 1990. This Baseline will be used in order to calculate the real effort of Kyoto, also considering the distribution of efforts among sectors.

Secondly, PRIMES provided the initial estimates of the "burden-sharing" among EU Member States. Thirdly, PRIMES – associated with POLES and GEM-E3 models – gave the main economic analysis highlighting the benefits of emission trading. It demonstrated the costs and benefits of trading, their allocation among sector and countries and the impact of different scenarios. The main conclusion – appended to Annex 1 of the 2000 Commission Green Paper on EU ETS – was that emission trading could reduce the cost of Kyoto by up to 30%. Looking at the Directive on a European Emission Trading Scheme (ETS) adopted in 2003, one can say that the PRIMES results were extremely useful in preparing the new EU position in terms of Climate Change policy, i.e. very favourable to the Kyoto flexible instruments, when it was not at all a EU position before Kyoto.

In the EU policy aimed at the promotion of renewable energy sources in Europe, the computer-based framework SAFIRE developed under the EU research activities (third EU RTD Framework Programme) at the beginning of the 1990s, provided the main renewable targets.

Firstly, SAFIRE provided the basis for setting the 1997 EU objective to double – from 6 to 12% – the share of renewables in the gross inland energy consumption by 2010 (Commission White Paper of 1997).

Secondly, SAFIRE provided the main figure on EU and national targets for the share of electricity produced from renewable energy sources. This will be formally acted in the proposal for a Directive and in the Directive on the promotion of renewable energy sources electricity, i.e. to increase "green electricity" from 14% to 22% by 2010. The results of the SAFIRE model were included in the Annex of the 2000 draft Directive.

In relation to energy efficiency policies in the EU, the electronic database MURE – developed under the second EU RTD Framework Programme – provided the basis for the main target of energy efficiency in EU, i.e. 18% by 2010. This figure was published in the energy efficiency Communication of 1998 and in the Action plan to improve energy efficiency in the European Community of 2000.

In the proposal for a Directive on the energy performance of buildings, the saving potential of 22% (heating, hot water, air-conditioning or lighting purposes) in buildings is also coming from the MURE database. In the 2003 proposal for a Directive and in the 2006 Directive on energy end-use efficiency, MURE (merged with the ODYSSEE database) provided the energy efficiency potential in Europe.

The 2005 Green Paper on energy efficiency stated that the EU could save at least 20% of its present energy consumption in a cost-effective manner, equivalent to € 60 Billion per year. This statement was coming from several scientific studies including those performed with MURE and ODYSSEE. All of the documents published in the Official Journal of the European Community/Union related to energy efficiency refer to MURE.

Concerning the measurement and internalisation of external costs in Europe, the EXTERNE methodology, sustained over the six EU RTD Framework Programmes, was the key reference in this field. The EXTERNE methodology and results have been widely used in the EU environmental policies.

EXTERNE was directly mentioned in three important policies. Firstly, EXTERNE was used to quantify the benefits of limiting certain pollutants into the air from large combustion plants. Close to ECU 4 Billion of benefits per year were mentioned. Secondly, EXTERNE was the main tool for the cost-benefit analysis of the Community strategy on air pollution. The figures of 60,000 avoided premature deaths and between € 42 and 135 Billion of health benefit per year were quoted in the Commission documents. Lastly, EXTERNE was the major reference in the Community guidelines on State aid for environmental protection where the renewable energy producer can receive up to € 0.05 per kWh due to their "avoided external costs".

Community framework for energy taxation

The principal role of the European Union in the field of taxation is to go towards a level of tax harmonisation sufficient to ensure that the full benefit of the internal market can be achieved. For this purpose, the EU and in particular the Commission works for a coherent and simple VAT system which takes account both of the budgetary needs of the Member States and which encourages economic growth. The Commission also aims to align the Member States' systems and rates of excise duties so as to remove economic distortions and thus to facilitate trade in the EU. In the field of direct taxation, the Commission is working to limit the economic distortions which result from different taxation systems. This action is concentrated on company and capital taxation

Although the first European scientific study demonstrating empirically the double dividend (positive for environment and positive for employment) using the HERMES model[2] was published in 1991, the 1992 Commission proposal for a tax on carbon dioxide (CO_2) emissions and energy never got consensus at the EU Council Minister of Finance and of Economic Affairs (ECOFIN)[3].

The political "blockage" remained with the 1995 amended CO_2/energy tax proposal[4]. In 1997, the Commission proposal seek to harmonise energy taxes by setting minimum levels of excise levies on all energy products[5]. This approach (setting Community minimum levels) was similar than the one followed in the 1992 Directive on the approximation of the rates of excise duties on mineral oils[6].

The Commission proposed that the Community minimum levels on mineral oils set by the 1992 Directive be up-rated and that minimum levels also be created for products other than mineral oils, i.e. essentially coal, natural gas and electricity. Some exemptions were offered to energy intensive industries and flexibility was allowed in respect of differentiating taxes according to environmental effects, as long as minimum rates were respected. *De facto*, this 1997 Commission proposal

---

2 European Commission, DG XII, *Increase of taxes of energy as a way to reduce CO_2 emissions: problems and accompanying measures*, Brussels, 1991. This study was written by V. Detemmerman, E. Donni and P. Zagamé.

3 European Commission, *Proposal for a Directive introducing a tax on carbon dioxide emissions and energy*, COM(92)226.

4 European Commission, *Amended proposal for a Council Directive introducing a tax on carbon dioxide emissions and energy*, COM(95)172.

5 European Commission, *Proposal for a Directive restructuring the Community framework for the taxation of energy products*, COM(97)30.

6 Council *Directive 92/81/EEC on the harmonization of the structures of excise duties on mineral oils*, OJ L 316, 31/10/1992 and Council *Directive 92/82/EEC on the approximation of the rates of excise duties on mineral oils*, OJ L 316, 31/10/1992.

suggested to extend the regime of minimum taxation levels on mineral oils by around 10 to 25% and introduced excises for other energy products.

The effects on national excise duty rates arising from the proposed increase in the EU minimum rate on unleaded petrol would have had an impact on three Member States (10% increase in Spain and 14% increase in both Greece and Luxembourg). The effects of EU minimum rate on diesel would have had an impact on seven Member States (3% increase in Austria and Belgium, 5% in Finland, 12% in Portugal, 14% in Spain, 18% in Luxembourg and 21% in Greece. The proposal Directive suggested to increase the minimum levels of indirect taxes (other than VAT) on motor fuels, heating fuels and electricity. For example, from ECU 287/1000 litres for unleaded petrol (motor fuel) in January 1997, the Directive proposed to increase it to ECU 417 in January 1998, ECU 450 in January 2000 and ECU 500 in January 2002.

In connection with this 1997 proposal, an "Assessment of the impact of the proposal" was published as a Working Paper of the Commission departments[7]. HERMES, GEM-E3 and E3ME, three macro-economic models including an energy module, built and up-dated under the EU RTD Framework Programmes, provided the evaluation of the impacts of the proposed EU energy taxation on economy, energy and environment[8]. The models provided inputs to the Community competitiveness questions and to the national preoccupations (GDP variation in the different Member States)[9].

Concretely, the findings published in the Commission Working Paper are coming from the results of scientists active in various European research projects in the energy socio-economic field in the EU RTD

---

7 European Commission, Working Paper of the Commission departments, *Presentation of the new Community system for the taxation of energy products*, Assessment of the impact of the proposal, SEC(97)1026/2.

8 European Commission, *Proposal for a Directive restructuring the Community framework for the taxation of energy products*, COM(97)30; European Commission, Staff working document, *Presentation of the new Community system for the taxation of energy products*, SEC(97)1026.

9 The HERMES model was also largely used by the Commission at the end of the eighties to evaluate the consequences of the completion of the European Internal Market. HERMES was developed by a network of European scientists from which the main authors were Alexander Italianer, Gonzales d'Alcantara and Paul Zagamé. Some other key actors in HERMES were Eric Donni and Francis Bossier. A. ITALIANER, G. D'ALCANTARA, P. ZAGAME, *The HERMES model for the Member States of the European Community in HERMES: Harmonised econometric research for modelling economic systems*, EC-North Holland, Luxembourg, 1993. Henry Guillaume and Jean-Michel Charpin used to say that HERMES was a major advancement in modelling due to the fact that two new concepts were developed: putty-clay approach for the production function and energy as a production factor.

Framework Programmes (see Chapter II). In particular, the following projects have to be mentioned: "Impacts of energy tax on the European economies" with a strong role of the Belgian Bureau Fédéral du Plan (Francis Bossier), "Climate technology strategy within competitive energy markets: towards new and sustainable growth" coordinated by the National Technical University of Athens (Pantelis Capros) and "A general energy-environmental-economy model for Europe – E3ME" coordinated by Cambridge Econometrics (Terry Barker)[10].

Table VI.1: Scientists results (HERMES, GEM-E3, E3ME models) used to assess the 1997 Commission proposal for a Community framework for the taxation of energy products (impacts of the proposal compared with the Baseline by the year 2005)

	HERMES	GEM-E3	E3ME
GDP	+ 0.06%	+ 0.02%	+ 0.20%
Employment EU15	+ 193,000	+ 155,000	+ 457,000
CO_2	- 1.7%	- 1.5%	- 0.5%

These modelling results (HERMES, GEM-E3, and E3ME) presented in the Commission Working Paper confirm that the 1997 proposal would have had a positive macroeconomic impacts if the tax revenues were used to reduce social security contributions paid by employers. For the EU as a whole, both GDP and employment were expected to be higher and CO_2 emissions were 0.5 to 1.7 percent lower. The positive EU-wide effects could be observed in practically all Member States. The sector impacts detailed in the Commission Working Paper were modest, with the energy sector expected to face the most negative impacts even if very limited (less than half percentage change from the Baseline).

Nevertheless, the 1997 proposed Directive did not reach the necessary unanimous support by the Member States in the Council negotiations. In the preparation of a new Directive, in 2002, a new study was carried out to update the Impact Assessment of the 1997 proposal[11]. The study was performed by the Catholic University of Leuven as coordinator (Denise van Regemorter). The tool used was the EU-wide general equilibrium model GEM-E3 developed within the energy socio-economic research of the fourth RTD Framework Programme, in particular the project entitled "European emission mitigation policy and technological evolution: economic evaluation with the GEM-E3-EG

10 R. BJERREGAARD, *Speech on environmental tax reform*, Conference on Environmental tax reform in Europe, Brussels, 27/10/1997.

11 Catholic University of Leuven, *Economic and environmental impacts of energy taxation*, 2003, Study for the Commission (DG Taxation and Customs Union).

model" coordinated by the German Centre for European Economic Research (ZEW, Tobias Schmidt)[12].

According to the study, taking into consideration the implementation of the EU minimum tax rates on energy products (€ 350/1,000 litres for professional diesel and € 360/1,000 litres for petrol and non-professional diesel), the macro-economic impacts would be negligible in the EU and the majority of the Member States. For a group of five countries (Austria, Belgium, France, Greece and Portugal) the impacts were slightly positive when the tax revenues were "recycled" to reduce the labour market contributions ("double dividend")[13].

Table VI.2: Scientists results (GEM-E3 model) used to assess the 2003 Directive restructuring the Community framework for the taxation of energy products (impacts of the 1997 revisited proposal compared with the Baseline by the year 2010)

	EU energy taxation with social security recycling compared to the Baseline	EU energy taxation with revenue for public deficit compared to the Baseline
GDP	0.00%	- 0.01%
Employment EU15	+ 0.02%	- 0.01%
CO_2	- 0.5%	- 0.54%

Nevertheless, as shown in Table IV.3, even if the impacts of this legislation were negligible in terms of costs, the Council was very modest in its 2003 Directive[14]. In fact, the level of the minimum taxation in the 2003 Directive was generally one third lower than the 1997 Commission proposal. In addition, these rates were below the existing rates in nearly half of the EU countries. In terms of timing, the 1997 proposal had a target of four years (i.e. by 2002) and the 2003 Directive had a target of six years (i.e. by 2010)[15].

---

12 The GEM-E3 computable general equilibrium model has also been used for illustrating the health related benefits of EU environmental policies (internalisation of external costs) and to develop estimates of the macroeconomic impacts of improved air quality within the European Commission *Clean Air for Europe programme (CAFÉ)*, COM(2001)245. See below the Chapter on "Measurement and internalisation of external costs in Europe".

13 D. PEARCE, *The role of carbon taxes in adjusting to global warming*, The Economic Journal, Vol. 101, 1991 and H. WELSCH, *Recycling of carbon/energy taxes and the labour market*, Environmental and resource economics, Vol. 8, 1996.

14 Council *Directive 2003/96/EC restructuring the Community framework for the taxation of energy products and electricity*, OJ, L 283, 27/10/2003.

15 European Commission, *Proposal for a Directive restructuring the Community framework for the taxation of energy products, op. cit.*; *Directive 2003/96/EC restructuring the Community framework for the taxation of energy products and electricity, op. cit.*

Table IV.3: Comparison of minimum levels of taxation between the 1997 Commission proposal and the 2003 Council Directive for motor fuels (petrol, kerosene, LPG, natural gas) and electricity

	1997 Commission proposal Target minimum levels of taxation for 1/1/2002	2003 Council Directive Minimum levels of taxation for 1/1/2010
Petrol (unleaded)	500 ECU/1000 l	359 €/1000 l
Kerosene	393 ECU/1000 l	330 €/1000 l
LPG	224 ECU/1000 kg	125 €/1000 kg
Natural gas	4.5 ECU/GJ	2.6 €/GJ
Electricity	3 ECU/MWh	0.5 €/MWh for business use 1 €/MWh for non business use

In relation to the impacts of fuel taxation on technology choice, one has to mention the PRIMES model used in the Annex of the Commission Green Paper on energy security of supply[16]. The PRIMES model was originally developed under the third EU RTD Framework Programme (project entitled "A computable price-driven equilibrium model of the energy system and markets for Europe – PRIMES" and coordinated by George Marsh from the UK Atomic Energy Authority). The study performed in the Green Paper was done by the National Technical University of Athens[17]. The study was largely coming from the results of the fourth RTD Framework Programme, project called "Climate technology strategy within competitive energy markets: towards new and sustainable growth" coordinated by the National Technical University of Athens (Pantelis Capros).

European Union climate change policy

Since the mid of the 1980s, the European Community (EC, later EU) has addressed Climate Change in a non-binding way[18]. In 1990, the EU committed itself to holding its CO_2 emissions in 2000 at or below their 1990 level[19]. The EU formalised the commitment when it signed the

16 European Commission, *Towards a European strategy for the security of energy supply*, COM(2000)769.

17 P. CAPROS, N. KOUVARITAKIS, L. MANTZOS, V. PANOS, L. VOUYOUKAS, *Note on the impact of fuel taxation on technology choice*, November 2000, Study for the Commission (DG Energy and Transport).

18 European Parliament, *Resolution on measures to combat rising CO_2 concentrations in the atmosphere*, OJ, C 255, 13/10/1986; Council, *Resolution on the greenhouse effect and the Community*, OJ, C 183, 21/6/1989.

19 European Commission, *Communication on Community policy targets on greenhouse issue*, SEC(90)496; Energy and Environmental Council, *EC climate change policy*, Luxembourg, 29/10/1990. A key European economic study published in 1991 will

United Nations Framework Convention on Climate Change (UNFCCC) on 3-14 June 1992 at the Rio Earth Summit (UN Conference on Environment and Development – UNCED). This position followed the Commission Staff Working Document and Communication of 1991 and 1992 on the "Community strategy to limit carbon dioxide emissions and improve energy efficiency"[20]. The main objectives and targets for the year 2000 were adopted in the 5th Environmental Action Programme adopted in March 1992[21].

In view of the third Conference of the Parties (COP-3) to the UNFCCC conference to be held in Kyoto from 1 to 11 December 1997, the EU had to prepare its negotiation position. The technical, economic and political feasibility of a 15% reduction of emissions from carbon dioxide (CO_2), methane (CH_4) and nitrous oxide (N_2O) by 2010 compared to 1990 for all industrialised countries, and in particular for the EU, were defined in the Commission Communication on "Climate Change – The EU approach for Kyoto" and in its connected Commission Staff Working Paper[22]. All the "Economic implications" of the European position before Kyoto (15% reduction of emissions for CO_2, CH_4 and N_2O in 2010 compared to 1990) included in this Communication were coming from the POLES and PRIMES models and the CRASH91 programme, all models and study developed by scientists under the RTD Framework Programmes.

The European research projects to be quoted (see Chapter II) are "Models for energy and environment: improvement and case studies (EFOM-ENV model)" coordinated by Dominique Gusbin from COHERENCE, "Data processing data banks" coordinated by Andrea

ensure the credibility of the models for climate change policy assessment in Europe for the next 15 years: European Commission, DG XII, *Cost-effectiveness analysis of CO_2 reduction options*, Brussels, 1991. This report was supervised by Pierre Valette and coordinated by Dominique Gusbin with the contributions of Ch. Houiller, P.E. Grohneit, P. Russ, T. Van Harmelen, F. Van Oostvoorn, J. Berry, Ch. Woods, G. Schmid, J. Psarras, W. Bocola, V. Martins, J. Seixas, L. Arimani and J.P. Garcia.

20 European Commission, *Community strategy to limit carbon dioxide emissions and improve energy efficiency*, SEC(91)1744; European Commission, *Community strategy to limit carbon dioxide emissions and improve energy efficiency*, COM(92)246. These initiatives have to be put in parallel with the taxation initiatives. See the Chapter on "Community framework for energy taxation" and more specifically the 1992 Commission proposal for a tax on carbon dioxide emissions and energy – COM(92)226.

21 European Commission, *Towards sustainability: A European Community programme of policy and action in relation to the environment and sustainable development*, COM(92)23.

22 European Commission, *Communication on climate change – The EU approach for Kyoto*, COM(97)481 and Commission Staff working paper, *Climate change – Analysis of proposed EU emission reduction objectives for Kyoto*, 1997.

Ricci from ISIS, "Development and applications of the HERMES model" coordinated by Jean-Eudes Moncomble from EURECO, "Finalisation of the CO_2 study/CRASH programme" coordinated by Frits van Oostvoorn from ECN and "Climate technology strategy within competitive energy markets: towards new and sustainable growth" coordinated by Pantelis Capros from the National Technical University of Athens. The so-called CRASH Programme will be a key document for the future economic analysis related to Climate Change. The coordinator of this study was Dominique Gusbin.

From the results of these models and studies, the Communication "Climate change – The EU approach for Kyoto" stated that:

> For a 15% reduction in CO_2 emissions compared to 1990, estimates of the direct compliance costs related to energy supply/demand mitigation actions range from around ECU 15 Billion to about ECU 35 Billion annually by 2010. This corresponds to roughly 0.2% and 0.4% of GDP in the year 2010[23].

Also in preparing its position for Kyoto, the main estimates of CO_2 emission trends were coming from the Commission Communication on "The energy dimension of climate change"[24]. The figures presented in this Communication were based on two studies performed by scientists with the MIDAS and later with the PRIMES energy models[25]. These studies were largely taking inputs from the European research results project "Climate technology strategy within competitive energy markets: towards new and sustainable growth" coordinated by the National Technical University of Athens (Pantelis Capros) and funded by the Fourth EU RTD Framework Programme.

The "pre-Kyoto scenario" (business as usual) of these studies provided the main reference figure for CO_2 emissions trends in the EU by 2010 that will be used in the future discussions and negotiations within the EU (among Member States) and in international fora like the UNFCCC conferences.

---

23 European Commission, *Communication on climate change – The EU approach for Kyoto*, *op. cit.*

24 European Commission, *The energy dimension of climate change*, COM(97)196.

25 National Technical University of Athens, *A business as usual energy scenario for pre-Kyoto*, April 1997, Study for the Commission (DG XVII–Energy); P. CAPROS, T. GEORGAKOPOULOS, L. MANTZOS, *Economic and energy system implications of European CO_2 mitigation strategy for 2010: a model based analysis*, National Technical University of Athens, 1997.

Table VI.4: Scientists results (PRIMES model) used by the EU in the Kyoto negotiations (CO_2 emissions trends in the pre-Kyoto scenario; and comparison between the modelling results and the final decision of the percentage of the Member States "burden-sharing")

Countries and EU	1990 ($MtCO_2$) EUROSTAT	2010 ($MtCO_2$) PRIMES	% change (1990-2010) "Baseline"	Pre-Kyoto target (CO_2) PRIMES	EU June 1998 burden-sharing (all GHG) EU Decision
AT	58	60	4.5%	-25%	-13%
BE	112	133	18.8%	-10%	-7.5%
DE	997	909	-8.8%	-25%	-21%
DK	53	59	9.6%	-25%	-21%
EL	72	103	43.7%	30%	25%
ES	209	289	38%	17%	15%
FI	54	74	36.9%	0%	0%
FR	367	389	5.9%	0%	0%
IE	31	39	24.9%	15%	13%
IT	401	475	18.3%	7%	-6.5%
NL	158	195	23.5%	-10%	-6%
PT	40	62	55.9%	43%	27%
SE	52	68	31.2%	5%	4%
UK	584	592	1.4%	-10%	-12.5%
EU	3188	3447	+ 8.1%	-10%	-8%

The figure of 8% increase of CO_2 emissions by 2010 in reference to 1990 in the EU is coming from the PRIMES model. This figure of 8% will be used in the different Communications mentioned above and in many others like in the 1999 Communication "Preparing for implementation of the Kyoto Protocol"[26]. This later was based on an updated study made with the same models and with scientists from National Technical University of Athens (Pantelis Capros), ESAP in Belgium (Jean-François Guilmot) and the CNRS-IEPE in France (Patrick Criqui and Dominique Finon)[27]. This study will also benefit from the so-called "Shared analysis" (coordinated by the German Fraunhofer Institute – Wolfgang Eichhammer) where national energy experts were associated to the Commission and scientists to define the assumptions, constraints and scenarios of the study.

[26] European Commission, *Preparing for implementation of the Kyoto Protocol*, COM(99)230.

[27] European Commission, *European Union energy outlook to 2020*, EC, Belgium, 1999.

Table VI.5 Scientists results (MIDAS and PRIMES models) used in the 1997 Commission Communications on Climate Change (sectoral distribution of energy-related CO_2 emissions)

Sector	1990 ($MtCO_2$)	2010 ($MtCO_2$)	% change (1990-2010)
Transport	743	1032	39%
Industry	626	532	-15%
Energy industry	141	158	12%
Domestic/Tertiary	654	680	4%
Elect./Heat production	1036	1057	2%
Total	3200	3459	+ 8.1%

The Kyoto Protocol was signed on 11 December 1997. The EU and its Member States committed to reduce total greenhouse gas (GHG) emissions by 8% in 2008 to 2012 relative to the year 1990[28]. The EU had firstly to find an agreement among its Member States. The "burden-sharing" was endorsed by the EU Environment Council on 16 June 1998. The burden-sharing agreement distributed the legally-binding Kyoto reduction target (-8%) over the EU Member States (Germany -21%, France 0%, Spain +15%, etc.). Secondly, the EU took the world leadership in the struggle against Climate Change. The EU made a strong diplomatic efforts (cf. negotiations with Russia, with emerging and developing countries, etc.) to get the Kyoto Protocol ratified, finally on 16 February 2005.

In order to achieve the Kyoto targets, the EU based its strategy on two pillars indicated in the Commission Communications "Climate Change – Towards a EU post-Kyoto strategy" and "Preparing for implementation of the Kyoto Protocol"[29]. These Communications dealt with the domestic policies and measures that were an integral part of the EU approach before Kyoto, and the so-called "flexible mechanisms" which were not part of the initial EU approach (before Kyoto). In fact, these flexible mechanisms were put forward during the Kyoto negotiations by the USA[30].

---

28 In the Kyoto Protocol, six greenhouse gas were covered: CO_2, CH_4, N_2O and HFC, SF_6 and PFC.

29 European Commission, *Communication on climate change – Towards a EU post-Kyoto strategy*, COM(98)353 and European Commission, *Preparing for implementation of the Kyoto Protocol*, COM(99)230.

30 Center for International Climate and Environmental Research (CICERO), *Climate change negotiations: Burden sharing and cost-effective implementation mechanisms and protocols – A conference summary*, TemaNord, 562, Denmark, 1997.

Figure VI.1: Scientists results (PRIMES model) used in the 1999 Commission Communication "Preparing for implementation of the Kyoto Protocol" (observed and projected greenhouse gases emissions for the EU)

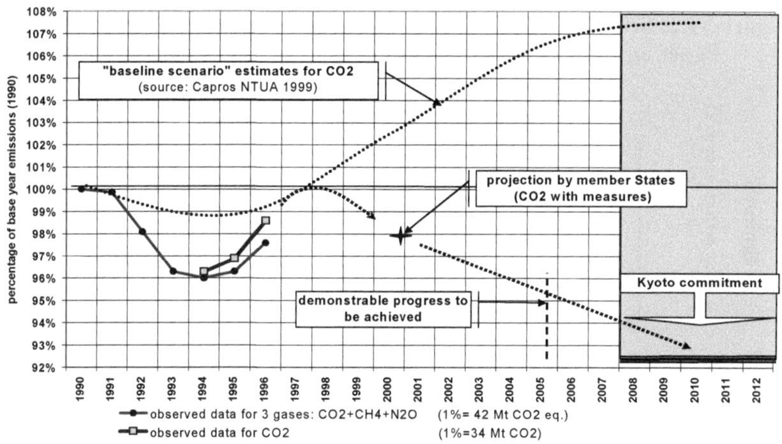

Following its two pillars Climate Change strategy (domestic policies and flexible instruments), the 8 March 2000, the European Commission presented two Communications. One concerned the "European Climate Change Programme (ECCP)" and the other the "Green Paper on greenhouse gas emissions trading within the EU"[31].

- The EECP Communication was quite short (12 pages) and with only a couple of footnotes (from almost twenty years, the EU dealt with common and coordinated policies in the EU, i.e. policies relevant to the household, industrial, commercial and transport sectors).
- The Green Paper on emissions trading was quite long (28 pages) and argued (48 footnotes)[32]. Of particular interest is the economic analysis mainly done with the European energy model PRIMES (and other models like GEM-E3 and POLES for the price of emis-

31 European Commission, *EU policies and measures to reduce greenhouse gas emissions: Towards a European climate change programme (ECCP)*, COM(2000)88 and European Commission, *Green Paper on greenhouse gas emissions trading within the European Union*, COM(2000)87.

32 In view of preparing this Green Paper, another scientific study – even if not related to modelling and to EU energy socio-economic activity – has to be mentioned in the preparation of the framework of the EU Emission Trading Scheme: Foundation for International Environmental Law and Development in London, *Designing options for implementing an Emissions Trading regime for greenhouse gases in the EC*, February 2000, Study for the Commission (DG Environment).

sion allowances)[33]. The modelling analysis concluded that emission trading at the EU level could reduce annual compliance cost of the Kyoto Protocol by as much as 30% (from € 9 to 6 Billion assuming that a national emission trading is in place in the Member States).

Figure VI.2: Scientists results (PRIMES model) used in the 2000 EU Green Paper on emission trading (costs of reaching the Kyoto target)

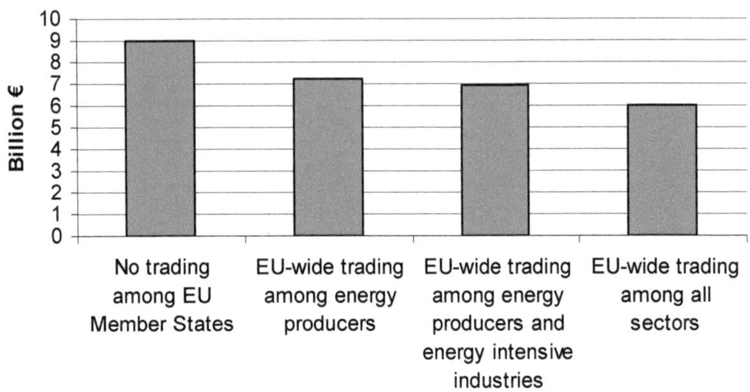

The PRIMES, POLES and GEM-E3 models provided also a comprehensive analysis of marginal abatement costs. In the absence of flexible instruments, the marginal abatement costs will differ widely between countries. The arithmetical average across countries was €60/tCO_2, but this covered a range from Germany with the lowest marginal cost (€25/tCO_2) to the Netherlands with the highest marginal cost (€160 /tCO_2). With full Annex B trading, the estimates of marginal costs fall within the range €10-30/tCO_2 around 2010[34].

In 2001, the Commission proposed a Directive on greenhouse gas emission allowance trading[35]. In the explanatory memorandum of this proposal, the Commission stated that:

---

33 National Technical University of Athens, *The economic effects of EU-wide industry-level emission trading to reduce greenhouse gases*, May 2000, Study for the Commission (DG Environment).

34 P. CAPROS *et al.*, *European environmental priorities: an environmental and economic assessment*, European Commission, DG Research, Conference on Climate negotiations and emission trading, Brussels, 29-30/08/2000. Results from P. CAPROS *et al.*, *Climate technology Strategy 1 – Controlling greenhouse gases, policy and technology options*, *op. cit.*

35 European Commission, *Proposal for a Directive establishing a scheme for greenhouse gas emission allowance trading within the Community and amending Council Directive 96/61/EC*, COM(2001)581. The Directive 96/61/EC concerns the Inte-

emissions reductions will be then made wherever in the Community it is cheapest to make them. The benefits of these cheaper reductions will be available to others elsewhere in the Community who may not themselves have as cheap reduction possibilities (...). The economic case for a Community-wide scheme is supported by several recent studies demonstrating efficiency gains.

Four studies, essentially based on the PRIMES and POLES energy models were quoted[36]. These studies were made by scientists and respectively coordinated by National Technical University of Athens (Pantelis Capros), Institute for Prospective Technological Studies in Sevilla – IPTS-Joint Research Centre (Antonio Soria and Peter Russ), ECOFYS in The Netherlands (Kornelis Blok) and COHERENCE in Belgium (Dominique Gusbin).

Following the co-decision procedure, i.e. two readings from the European Parliament and the common position from the Council (See Chapter VIII), an amended proposal for a Directive was prepared by the Commission at the end of 2002[37]. The greenhouse gas emissions trading Directive was finally adopted in October 2003[38]. The Directive limits the total carbon dioxide emissions from almost 12000 installations across the EU's 25 Member States (EU-25). From January 2005, the European Union greenhouse gas Emission Trading Scheme (ETS) commenced operation as the largest multi-country, multi-sector greenhouse as ETS world-wide[39].

---

grated Pollution Prevention and Control (IPPC Directive). This IPPC Directive is the basis of many European environment legislation like the Commission *proposal for a Directive on the geological storage of carbon dioxide*, COM(2008)18.

36 European Commission (study for DG Environment), *The economic effects of EU-wide industry-level emission trading to reduce greenhouse gases – Results from PRIMES model*, May 2000; European Commission (study for DG Environment), *Preliminary analysis of the implementation of a EU-wide permit system on CO_2 Emissions abatement costs – Results from POLES model*, April 2000; European Commission (study for DG Environment), *Economic evaluation of sectorial emission reduction objectives for climate change*, March 2001; and European Commission (study for DG Environment), *Economic evaluation of quantitative objectives for climate change*, 1999.

37 European Commission, *Amended proposal for a Directive establishing a scheme for greenhouse gas emission allowance trading within the Community and amending Council Directive 96/61/EC*, COM(2002)680.

38 European Parliament and Council *Directive 2003/87/EC establishing a scheme for greenhouse gas emission allowance trading within the Community and amending Council Directive 96/61/EC*, OJ, L 275, 25/10/2003.

39 Although the EU CO_2 ETS is the world's largest, the US has the distinction of having provided the seedbed for public policy experiments with emissions trading. In fact, the US SO_2 trading program ("Acid Rain Program") started in 1995. D. ELLERMAN, R. SCHMALENSEE, P.L. JOSKOW, J.-P. MONTERO, E. BAILEY, *Markets for*

For the future EU climate policy, many of the EU policies that are already in place will have an important impact beyond the Kyoto Protocol's first commitment period. The EU greenhouse gas ETS automatically continues in 5-year periods after 2012. The second phase of the European Climate Change Programme has been launched at the end of 2005 (ECCP) and will be expanded to include carbon capture and storage, passenger road transport, aviation and strategies to adapt to the effects of climate change, in addition to the work programme already covered. The European Commission has also adopted a Communication outlining key elements for a strategy for post-2012 actions[40].

This 2005 Communication indicates that the EU is ready to engage in an open dialogue between countries concerning the further development of an international framework post-2012, it has highlighted a number of key elements for a successful global climate policy. These include the need for broader participation by countries and sectors, the development of low-carbon technologies, the continued and expanded use of market-based instruments, and the need to adapt to the inevitable impacts of Climate Change. The Commission will review progress and explore new actions systematically to exploit cost effective emission reduction options in synergy with the Lisbon strategy.

The figures and the political messages indicated in the 2005 Commission Communication "Winning the battle against global Climate Change" and especially its Staff Working Paper (Annex to the Communication) were largely based on the results of scientists and modellers[41]. The "Greenhouse gas Reduction Pathways" (GRP) study was coordinated by the CNRS/LEPII (Patrick Criqui) using the POLES model, with three main partners: RIVM/MNP (Michel Den Elzen and Marcel Berk), ICCS-NTUA (Pantelis Capros and Nikos Kouvaritakis) and CES-KUL (Denise Van Regemorter).

The 2005 Commission Communication sets out key elements for the EU's climate strategy among which the full use of the Kyoto market-based instruments, the largest possible participation of countries in the world, the inclusion of more sectors and gases, the deployment of new energy and transport technologies and the importance to work on "adaptation".

---

clean air: the US acid rain program, Cambridge University Press, Cambridge and New York, 2000.

40 European Commission, *Winning the battle against global climate change*, COM(2005)35 and European Commission, Staff working paper, *Winning the battle against global climate change*, SEC(2005)180.

41 European Commission (study for DG Environment), *Greenhouse gas reduction pathways in the UNFCCC process up to 2025*, October 2003.

The greenhouse gas emissions reduction targets for 2020 taken by the Spring 2007 European Council (Heads of States and Governments of the EU) are based on the Commission Communication "Limiting global climate change to 2 degrees Celsius – The way ahead for 2020 and beyond"[42].

This Communication proposes that:

> The EU pursues in the context of international negotiations the objective of 30% reduction in greenhouse gas emissions (GHG) by developed countries by 2020 (compared to 1990 levels). This is necessary to ensure that the world stays within the 2°C limit. Until an international agreement is concluded, and without prejudice to its position in international negotiations, the EU should already now take on a firm independent commitment to achieve at least a 20% reduction of GHG emissions by 2020, by the EU emission trading scheme (EU ETS), other climate change policies and actions in the context of the energy policy. This approach will allow the EU to demonstrate international leadership on climate issues. It will also give a signal to industry that the ETS will continue beyond 2012 and will encourage investment in emission reduction technologies and low carbon alternatives. By 2050 global emissions must be reduced by up to 50% compared to 1990, implying reductions in developed countries of 60-80% by 2050. Many developing countries will also need to significantly reduce their emissions.

This Communication is scientifically supported by its Impact Assessment (Commission Staff Working document accompanying the Communication)[43]. The economic analysis has been done mainly by the Joint Research Centre – IPTS (Antonio Soria and Peter Russ) using the POLES and GEM-E3 models.

The envisaged reduction scenario aims to meet the objective of preventing an average global temperature increase of more than 2°C. For that purpose, the CO_2 equivalent concentration levels need to be reduced to 450 ppmv in the long-term, i.e. reaching 20 Gt CO_2 by 2050 against more than 50 Gt CO_2 in the Baseline.

Being done with the POLES model, this reduction scenario shows a EU reduction of almost 60% by 2050 and of 21% by 2020. At the European Council Summit on 8-9 March 2007, the Heads of State and Government will decide that "the EU makes a firm independent commitment to achieve at least a 20% reduction of greenhouse gas emissions by 2020 compared to 1990"[44].

---

42 European Commission, *Limiting global climate change to 2 degrees Celsius – The way ahead for 2020 and beyond*, COM(2007)2.

43 European Commission, *Staff working document accompanying the Communication Limiting global climate change to 2 degrees Celsius – The way ahead for 2020* and beyond, SEC(2007)8.

44 European Council, *Conclusions*, 8-9 March 2007.

Figure VI.3: Scientists results (POLES model) used in the 2007 Commission Communication and by the Spring 2007 European Council (relative development of greenhouse gas emissions in the EU-25 in the baseline and in the reduction scenario)

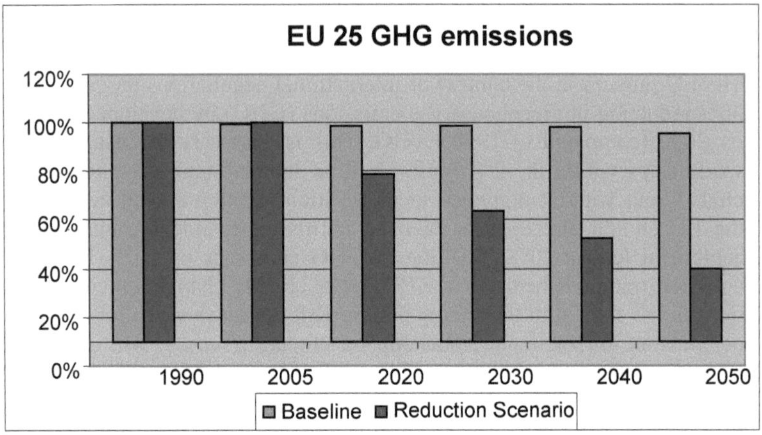

Promotion of renewable energy sources in Europe

In its 1995 White Paper "An energy policy for the European Union", the Commission put forward its views as regards Community energy policy objectives and instruments to achieve them. Three policy objectives were identified: improved competitiveness, security of supply and protection of the environment. The promotion of renewables was seen as an important factor to achieve these objectives[45].

In 1996 and 1997, a complete analysis of the challenges for the renewables sector, including the current and expected renewables production in EU Member States was published in the Commission Green and White Papers[46]. In these two Communications, the ambitious indicative target of 12% – a doubling – of renewable sources of energy in the EU gross inland energy consumption by 2010 was presented.

45 European Commission, *An energy policy for the European Union*, COM(95)682.

46 European Commission, *Energy for the future: Renewable sources of energy*, COM(96)576 and European Commission, *Energy for the future: Renewable sources of energy*, COM(97)599.

Table VI.6: Scientists results (SAFIRE model in the TERES study) used in the EU 1997 White Paper on renewable sources of energy (expected contribution of renewable sources and technologies)

Renewable Energy Technologies(RET)	1995	2010
For ELECTRICITY	(in GWh)	(in GWh)
Wind	4,169	82366
Small Hydro	34,314	60701
Large Hydro	273,577	280,540
Photovoltaic	49	12,213
Geothermal	4,405	10,850
Solar thermal	0	3,504
Tide	542	4,883
Landfill gas	1,351	10,904
Municipal waste	7,900	19,231
Industrial waste	12,249	19,432
Agricultural waste	133	500
Wood crops	772	42,939
Forest residues	2,753	4,387
For HEAT	(in ktoe)	(in ktoe)
Landfill gas	85	2,634
Municipal waste	2,580	4,202
Industrial waste	7,692	18,535
Agricultural waste	245	3,371
Wood crops	548	25,666
Forest residues	20,829	27,986
Geothermal	628	8,111
Solar thermal	287	5,089
For TRANSPORT	(in ktoe)	(in ktoe)
Ethanol/Biodiesel	168	15,001
RET ENERGY	73 Mtoe	197 Mtoe
PRIMARY ENERGY DEMAND	1,394	1,571
% RET contribution	5.2%	12.5%

This 12% target was coming from the TERES study performed with the SAFIRE model[47]. The SAFIRE model was developed by the UK Energy for Sustainable Development team (ESD – Gareth Jones) in the framework of the third EU RTD Framework Programme, project

47 European Commission (study for DG XVII-Energy), *TERES and TERES II, The European Renewable Energy Study, The prospects for renewable energy in 30 European Countries from 1995-2020*, November 1996, ALTENER programme.

"SOLFEGE-TEG-SAFIRE-SITRE models" coordinated by the Catholic University of Louvain (Yves Smeers). The main researchers that performed the TERES studies were from Energy for Sustainable Development – ESD (Ian McChesney, Mike Bess and Mark Whiteley).

Concretely, the assessments country-by-country produced from SAFIRE, combined with the physical assessments from a variety of technical projects, form the basis for the 12% objective of the White Paper.

In 2000, the proposal for a Directive on the promotion of electricity from renewable energy sources included an indicative target to increase the proportion of the EU-15 electricity to be supplied by renewable sources to reach 22% in 2010 (against 14% in 1997). The draft Directive explicitly mentioned that, "the energy module used is SAFIRE which was used already in the TERES II study and was originally developed under the JOULE II programme"[48].

In 2001, the European Parliament and the Council adopted the Directive on the promotion of electricity produced from renewable energy sources in the internal electricity market[49]. The EU target for green electricity remained the same at 22%. Twelve Member States out of fifteen took exactly the same figures than in the draft Directive, i.e. those coming from the SAFIRE model. Finland, Portugal and The Netherlands reduced their renewable energy sources electricity target in the Directive respectively from 35 to 31.5%, from 45.6 to 39% and from 12 to 9%.

In addition to setting targets for renewable electricity, the Directive allows the coexistence of various different national support schemes, expects that Member States ensure the guarantee of origin of electricity produced from renewable energy sources, reduce the administrative procedures and take the necessary measures to ensure the transmission and distribution of electricity produced from renewable energy sources.

---

48 European Commission, *Proposal for a Directive on the promotion of electricity produced from renewable energy sources in the internal electricity market*, COM(2000)279. The JOULE II programme was the non nuclear energy programme of the third EU RTD Framework Programme (second Call for Proposals) See the Chapter of this book on "Energy research in the European Union".

49 European Parliament and Council *Directive 2001/77/EC on the promotion of electricity produced from renewable energy sources in the internal electricity market*, OJ, L 283, 27/10/2001.

Table VI.7: Scientists results (SAFIRE model) used in the proposal and in the 2001 Directive on electricity produced from renewable energy sources (percentage of green electricity)

	RES-E% 1997	RES-E% 2010	RES-E% 1997 without large hydro	RES-E% 2010 without large hydro
Austria	72.7	78.1	10.7	21.1
Belgium	1.1	6	0.9	5.8
Denmark	8.7	29	8.7	29.0
Finland	24.7	35 (31.5)*	10.4	21.7
France	15.0	21	2.2	8.9
Germany	4.5	12.5	2.4	10.3
Greece	8.6	20.1	0.4	14.5
Ireland	3.6	13.2	1.1	11.7
Italy	16.0	25	4.5	14.9
Luxembourg	2.1	5.7	2.1	5.7
Netherlands	3.5	12 (9)*	3.5	12.0
Portugal	38.5	45.6 (39)*	4.8	21.5
Spain	19.9	29.4	3.6	17.5
Sweden	49.1	60	5.1	15.7
UK	1.7	10	0.9	9.3
European Union	13.9%	22.1%	3.2	12.5%

* The figures between parentheses are those indicated in the Directive.

Even if not "modelling-oriented", another EU research project funded by the fourth RTD Framework Programme has to be mentioned as it was an exemplary project in terms of involvement of stakeholders. Called "Renewable electricity in liberalising markets" and coordinated by Energy for Sustainable Development (ESD – Mike Bess), this project played an important role in involving some major stakeholders in the debate around the Directive on green electricity. The 1998-1999 research was designed to feed into the contemporary process to develop a working paper on "Electricity from renewable energy sources and the internal electricity market", with the clear intention to provide inputs to the future Directive. Among the conclusions of this project, one can read:

- Overall (binding) target for growth in renewable energy would need to be adopted at the level of the Union;
- A EU renewables "bubble" is desirable, within which individual Member States would negotiate a (binding) target for renewables growth;
- Certification of the origin of all renewable electricity is a crucial condition for this to work;

- It is necessary to develop a single EU-wide unified procedure for the certification of origin;
- Certificates should be traded between market players and between member states independently of physical power flows.

The results of the project were widely disseminated, using fully the access of large industrial partners to government in a policy area that was to them of legitimate concern. This permitted discussion among key stakeholders in the power sector including government policy makers, and NGOs. This project initiated the idea of "RECS", the European Renewable Energy Certificate System. A RECS Group was formed, comprising organisations from Austria, Denmark, France, Germany, Italy, Norway, The Netherlands and UK and has become a focus for the development of an independent, industry-led, European-wide renewable certificate trading system.

Many of the ideas generated in this project can be detected in the draft Directive on renewables and in the explanatory memorandum even if the draft Directive went less far than the proposals made in the project. Nevertheless, these ideas on a European wide scheme of certification and support could maybe find an expression in the future.

Concerning the promotion of renewables, in addition to the green electricity Directive, one has to mention the Directive on the promotion of biofuels that was adopted in 2003 and the one on cogeneration ("CHP Directive") at the beginning of 2004[50]. According to the biofuels Directive, 5.75% of transport fuel should consist of biofuels by 2010 (indicative target). The CHP Directive aims at promotion and development of high efficient cogeneration units; it calls for development of support schemes in the Member States based on the useful heat demand and primary energy savings and it establishes a framework for a scheme for a guarantee of origin of CHP electricity.

Also interesting renewable energy sources development in Europe, the 2002 Directive on energy performance of buildings (See below) and the 2003 Directive on the taxation of energy products (See above) have to be reminded here.

In 2004, the Commission published a Communication and a Staff Working Document on "The share of renewable energy in the EU"[51]. The analysis showed that with current policies and measures, 18% – rather than the 22% green electricity target – will be achieved by 2010.

50 European Parliament and Council *Directive 2003/30/EC on the promotion of biofuels*, OJ, L 123, 17/5/2003; European Parliament and Council *Directive 2004/8/EC on the promotion of cogeneration*, OJ, L 52, 21/2/2004.

51 European Commission, Staff working document, *The share of renewable energy in the EU*, COM(2004)366 and European Commission, SEC(2004)547.

In the Commission Staff Working Document, a reference is already made at the evolution of renewable energy sources by 2020. The study FORRES 2020 coordinated by the German Fraunhofer Institute (Mario Ragwitz) is mentioned[52]. FORRES provided the main data for the penetration, potentials and costs of renewable energy sources for electricity, heat and biofuels by 2020. FORRES used the model GREEN-X developed by the Technical University of Vienna (Reinhard Haas) in the Fifth RTD Framework Programme, project called "Deriving optimal promotion for increasing the share of renewable energy sources electricity in a dynamic European electricity market". GREEN-X model (and PRIMES) will also the basis if the Impact Assessment of the EU renewable objective (20% by 2020) adopted by the Council in Spring 2007[53].

Energy efficiency in the European Community

Since the first oil crisis in 1973, energy efficiency and energy saving were a stated policy objective of the Community and its Member States in order to reduce oil imports and improve security of supply. In general terms, energy efficiency means obtaining more energy services from less primary energy; energy saving refers to the reduction of unnecessary energy use[54]. Both energy efficiency and energy saving follow the same long-term strategy: a reduction of energy demand.

Energy efficiency and energy saving resulted directly from the higher energy prices and the expectation that they would further increase. As these pressures disappeared, so has much of the effort to improve energy efficiency. If the energy intensity in the EU has continuously improved these last 30 years, after 1986, when energy prices fell sharply, energy efficiency improvements continued but at a much slower pace especially during the 1990s[55].

In 1998, a new Commission Communication on energy efficiency was published[56]. Energy efficiency applies to a number of activities

52 M. RAGWITZ, J. SCHLEICH, C. HUBER, G. RESCH, T. FABER, M. VOOGT, R. COENRAADS, H. CLEIJNE, P. BONDO, *Analyses of the EU renewable energy sources evolution up to 2020 (FORRES 2020)*, April 2005, Study for the Commission (DG Energy and Transport).

53 European Commission, Staff working document, *Renewable energy road map – Renewable energies in the 21st century: building a more sustainable future*, SEC (2006)1719.

54 For the energy economy as a whole, energy intensity is the gross inland energy consumption divided by GDP.

55 B. LAPONCHE, B. JAMET, M. COLOMBIER, S. ATTALI, *Maîtrise de l'énergie pour un monde vivable*, ICE, Paris, 1997.

56 European Commission, *Energy efficiency in the European Community – towards a strategy for the rational use of energy*, COM(98)246.

related to the production and distribution of energy to final energy as well as use or end-use. One of the main objectives of this Communication was to look at the potential for energy efficiency until 2010. For this purpose, the Communication focused on realising as far as possible the "economic potential" for improving energy efficiency.

The economic potential is the potential saving which can be achieved by optimising costs and choosing the best available cost-effective technology, assuming that rational requirements for pay-back periods and rates of return on investment are met. Economic potential is larger than "market potential" (i.e. business as usual energy saving that can be expected to be achieved in practice by individuals and firms taking into account the energy prices and the degree of market barriers, due for instance, to imperfect knowledge). Achieving the economic potential normally requires the removal of market barriers. The price of energy plays the same important role as with market potential[57]. Due to the natural, cost-effective rate of replacement for household appliances, lighting, buildings, industrial processes, motors, vehicles, etc., the economic potential is very much a function of time. Technology is not replaced before necessary. It is replaced as it wears out or becomes uneconomical.

The economic potential can best be calculated by using a Bottom-up method wherein the total current stock of energy-using equipment in the EU is modelled, along with the energy consumption this stock requires. Future stock growth and stock replacement are then calculated and extrapolated. Two scenarios are developed: the business as usual ("Baseline"), i.e. market rate of replacement and choice of technology; and the scenario which is calculated using the same rates of growth as in the Baseline but assuming gradual removal of market barriers and greater use of cost-effective energy-efficient technology.

The rate of market penetration of efficient technology is based on historical trends adjusted for increasing market shares. The difference between the two calculations defines the economic potential.

The MURE electronic database was used for such calculations. MURE was developed originally by European researchers under the second EU RTD Framework Programme, under the projects "Data processing data banks" and "MURE, COST, Biomass project completition" coordinated by the Italian Instituto di Studi per l'Informatica e i

[57] In addition to the economic and market potentials, the "societal potential" refers to the potential which can be achieved if further investments in energy efficiency are undertaken based on what is cost-effective for society as a whole including e.g. security of supply and sustainability; and the "technical potential" represents the upper limit of energy savings as it assumes the use of the most energy-efficient technology available, regardless of whether it is cost-effective or not.

Sistemi (ISIS – Andrea Ricci). The calculations for the 1998 Communication were made with MURE through the DG Energy SAVE programme (non technological programme for energy efficiency) and with other European energy specialised organisations like ADEME (Didier Bosseboeuf) and ENERDATA in France (Bertrand Chateau and Bruno Lapillonne) and the German Fraunhofer Institute (Wolfgang Eichhammer)[58].

Table VI.8: Scientists results (MURE database) used in the 1998 Commission Communication on energy efficiency (available economic potential in the EU by 2010)

Sector	Final energy consumption 1995	Available potential to 2010 (% of 1995 consumption)
Industry	268 Mtoe	17%
Transport	270 Mtoe	14%
Domestic and tertiary	358 Mtoe	22%
Total	896 Mtoe	18%

The 1998 energy efficiency Communication explicitly stated: "Estimates are that the available economic potential between 1998 and the year 2010, for all sectors combined, is approximately 18% of final annual energy consumption"[59]. This is calculated based on final annual energy consumption in 1995, and assumes a market or autonomous improvement in energy intensity of 0.6% per year, which was the trend of improvement between 1990 and 1995. Energy prices are assumed to remain constant.

As indicated in the above-mentioned table, while there is considerable potential in all sectors, domestic and tertiary remains the largest. Industry, especially the energy-intensive industry, has carried out many efficiency measures[60].

In 2000, the "Action plan to improve energy efficiency in the European Community" made a clear reference to MURE. The same target of 18% by 2010 is mentioned.

> The underlying assumption behind a plan of action for energy efficiency is that, whilst market forces have gradually improved energy efficiency in past years, it is clear that additional steps can and should be taken. There still

---

[58] European Commission (study for DG XVII-Energy), *MURE model estimation on energy efficiency in the EC*, March 1998, SAVE programme.

[59] European Commission, *Energy efficiency in the European Community – towards a strategy for the rational use of energy*, *op. cit.*

[60] The calculation for the transport sector is based entirely on fuel efficiency improvements. If other effects, including modal shifts, are included, the potential for the transport sector is estimated to be around 25% by the year 2010.

remains a large economic potential for further improvement that can be estimated now to be over 18% of present energy consumption[61].

The household and tertiary sector (160 Million buildings in the EU-15) is responsible for approximately 40% of the final energy demand (heating represents 70% of domestic energy consumption and 50% of commercial energy consumption) in the EU and has the highest potential for saving. In 2001, a proposal for a Directive on the energy performance of buildings was presented by the Commission. The 22% energy saving potential in buildings coming from the MURE database was explicitly mentioned in the proposal. The draft Directive stated:

> As regards energy in buildings that is used for heating, hot water, air-conditioning or lighting purposes, a saving potential of around 22% of present consumption is estimated to exist and can be realised by the year 2010[62].

A Directive on energy performance of buildings was adopted in 2003[63]. It deals with the integrated energy performance standards, application of these standards to new buildings and to certain existing buildings when they are renovated, certification schemes and specific inspection and assessment of boilers and heating/cooling installations.

At the end of 2003, a proposal for a Directive on energy end-use efficiency and energy services was presented by the Commission. The purpose of this Directive was to enhance the cost-effective and efficient end-use of energy in the Member States. Firstly, it aimed to provide the necessary targets, mechanisms, incentives and institutional, financial and legal frameworks to remove existing market barriers and imperfections for the efficient use of energy. Secondly, it aimed to develop a market for energy services and for the delivery of energy efficiency programmes and other energy efficiency measures to end users. This proposal for a Directive dealt, among others, with the procurement of energy end-use efficiency by the public sector, financial instruments for energy savings and energy audits. Again, the MURE database was quoted in order to give the energy efficiency potential:

> For industry this potential is estimated to be approximately 17% of current final consumption, realisable by 2010. For the domestic and tertiary sector, it is 22% and for transport 14%, excluding modal shifts[64].

---

61 European Commission, *Action plan to improve energy efficiency in the European Community*, COM(2000)247.

62 European Commission, *Proposal for a Directive on the energy performance of buildings*, COM(2001)226.

63 European Parliament and Council *Directive 2002/91/EC on energy performance of buildings*, OJ, L1, 4/1/2003.

64 European Commission, *Proposal for a Directive on energy end-use efficiency and energy services*, COM(2003)739.

In 2005, a Green Paper on energy efficiency was published. It stated:

That the EU could save at least 20% of its present energy consumption in a cost-effective manner, equivalent to € 60 Billion per year, or the present combined energy consumption of Germany and Finland[65].

Several studies are quoted and among others, those from MURE and ODYSSEE led by ISIS in Italy (Stefano Faberi and Andrea Ricci) and ADEME in France (Didier Bosseboeuf). In addition to several references to expert groups, the Spring 2007 European Council will finally adopt the 20% energy efficiency improvement targets (in relation to the baseline)[66].

The Green Paper highlighted the "negajoules" defined as the energy savings calculated on the basis of 1971 energy intensity. *De facto*, in the EU-27 the energy demand is today of 1700 Mtoe while it would have been of 2700 Mtoe without progresses in energy efficiency.

Figure VI.4: Development of the primary energy demand and of "negajoules" in the European Union – EU25 mentioned in the Commission 2005 Green Paper on energy efficiency

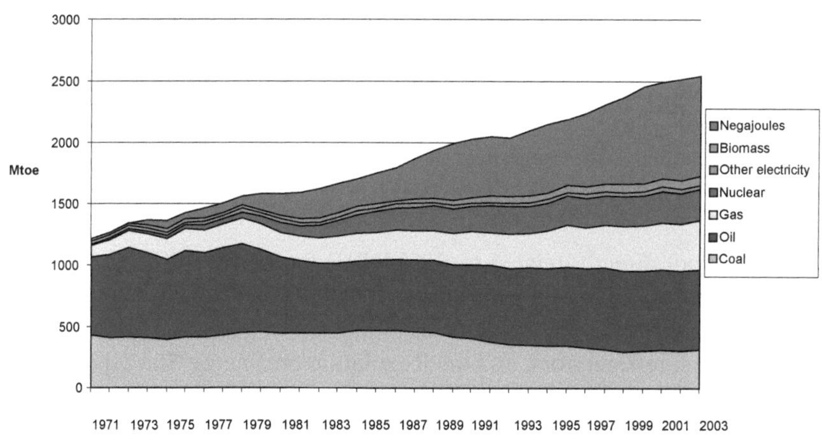

At the world level, the impact of energy efficiency is still higher. According to a UN Foundation study, since 1990, energy efficiency allowed to cover 52% of the new energy services demands while new energy supplies have contributed 48% of new energy service demands.

65 European Commission, *Green Paper on energy efficiency or doing more with less*, COM(2005)265.

66 European Commission, *Action Plan for energy efficiency: Realising the potential*, COM(2006)545; European Commission, Staff working document, *Action Plan for energy efficiency: Realising the potential*, SEC (2006)1173.

Without energy efficiency investments (broadly defined as the difference between the 1990 and 2006 energy intensities), worldwide energy use in 2006 would have been twice the 1990 level[67]. The 366 Exajoules (EJ) used in 1990 would have been 612 EJ in 2006 with the 1990 technology and market structure. In reality, thanks to new energy efficient technologies and energy efficiency investments since 1990, the world energy demand is of 485 EJ in 2006.

In the 2006 adopted Directive on energy end-use efficiency, the MURE electronic database (merged with the ODYSSEE one) was not only mentioned but even suggested as a major methodology to be used. In its Annex IV, the Directive stated:

> In measuring the realised energy savings (...), a harmonised calculation model (...) shall be used. A Top-down calculation method means that the amount of energy savings is calculated using the national or larger-scale aggregated sectoral levels of energy savings as the starting point. Adjustments of the annual data are then made for extraneous factors such as degree-days, structural changes, product mix, etc. to derive a measure that gives a fair indication of total energy efficiency improvements. This method does not provide exact measurements at a detailed level nor does it show cause and effect relationships between measures and their resulting energy savings. However, it is usually simpler and less costly and is often referred to as "energy efficiency indicators" because it gives an indication of developments. In developing the Top-down calculation method used in this harmonised calculation model, the Committee shall base its work, to the extent possible, on existing methodologies such as the ODEX model (i.e. ODYSSEE-MURE)[68].

Even if not directly related to the MURE (and ODYSSE) database, concerning energy efficiency in Europe, it is worthwhile mentioning:

The various Directives on labelling of electric ovens, air-conditionners, refrigerators, and the Regulation on Energy Star labelling for office equipment[69];

- the "Energy-using Products" (EuP) proposal for a Directive aimed to reduce the impacts in the product design phase of heating, electric motors, lighting and domestic appliances. This proposal was presented by the Commission in 2003 and adopted by the European Parliament and Council in 2005[70]. This Directive paves the way for

67 United Nations Foundation – Expert group on energy efficiency, *Realizing the potential of energy efficiency*, Washington, 2007.

68 European Parliament and Council *Directive 2006/32/EC on energy end-use efficiency and energy services and repealing Directive 93/76/EEC*, OJ, L 114, 27/4/2006.

69 Directives 2002/40/EC; 2002/31/EC; 2003/36/EC; Regulation 2422/2001/EC.

70 European Commission, *Proposal for a Directive on establishing a framework for the setting of Eco-design requirements for Energy-Using Products and amending*

the definition of eco-design specifications on a product-specific basis. Concretely, the design of products such as hairdryers, fridges, computers, office equipment, etc. should be more environmentally-friendly and reduce emissions of greenhouse gases.

EU Measurement and internalisation of external costs

External costs are the costs imposed on society and the environment that are not accounted for by producers and consumers of energy, i.e. that are not included in the market (private) price. They reveal, for example for the electricity production or for transport sector, the fuels and technologies that have the largest impacts (damages) on health and environment[71].

Since 1988, empirical attempts to derive external or social costs have mostly concentrated on the energy sector as one of the major contributors to environmental problems. At the same time, some work has also been done on the transport sector, as another major source of pollution and negative health impacts[72].

One of the major objectives of the EXTERNE energy socio-economic research project was to quantify the energy (production and consumption) damages to human health and environment coming from air pollution. The EXTERNE research work started with a collaboration between EU and USA teams in the second RTD Framework Programme, projects "External costs of fuel cycle" and "Scientific advice for the EC/US joint study on fuel cycle costs" in 1991 coordinated respectively by Metroeconomica (Anil Markandya) and AEA Technology (George Marsh) in the UK.

From 1992 to 1998, an effort was made to consolidate the external costs methodology in the third EU RTD Framework Programme and research was oriented towards national case studies and electricity/transport sectors in the fourth EU RTD Framework Programme. Two major projects have to be mentioned: "National implementation in the EU of the EXTERNE accounting framework" coordinated by the Spanish CIEMAT (Julio Montes) and the project External costs of energy

Council Directive 92/42/EC, COM(2003)453; and European Parliament and Council *Directive 2005/32/EC establishing a framework for the setting of eco-design requirements for energy-using products and amending Directives 92/42/EEC, 96/57/EC and 2000/55/EC*, OJ, L191, 22/7/2005.

71 D. ROSSETTI di VALDALBERO, J. BALEN, *Energy and transport: pricing externalities in the European Union – Setting concepts into motion: sustainable development and R&D policies*, OSTC, Proceedings, Brussels, 2001.

72 O. HOHMEYER, R.L. OTTINGER, *Social costs of energy*, Springer-Verlag, Berlin, 1994.

conversion – improvement of the EXTERNE methodology and assessment of energy-related transport externalities coordinated by the University of Stuttgart-IER (Rainer Friedrich). These two projects succeed to involve all the main research teams of Europe working in this field like – in addition to those already mentioned – VITO in Belgium, Risoe in Denmark, École des Mines in France, EKONO and VTT in Finland, Fondazione ENI and Bocconi-IEFE in Italy, University College of Dublin in Ireland, Vrije Universiteit – IVM in Amsterdam and CEETA in Portugal[73].

Politically speaking, the concept of internalisation of external costs in Europe dates back from the end of the 1980s. Some important European and Commission documents explicitly refer to this concept.

Although the first Community environmental action programme dates back to 1973, legally speaking, the Community environmental policy has been introduced in 1986 with the article 130 R of the Single European Act (SEA). The polluter-pays principle and the integration of the environmental concerns in the other European policies are explicitly recognised. The same article of the SEA stipulates that the Community, in preparing its policy on environment will, among others factors, take account of available scientific and technical data and, more directly related to the concept of external costs, will take account of the potential benefits and costs of action or lack of action.

De facto, it is the *cost-benefit analysis* that is mentioned. This kind of analysis aims to compare all the advantages in relation to their costs. If the benefit/cost ratio is higher than 1, then the envisaged policy or programme has a social value. These cost-benefit analyses have been largely promoted by environmental economists that look at the relationship between economic growth, environmental pollution and social well-being.

These analyses have acquired strengths in political discourse thanks to authors like David Pearce and Anil Markandya in the UK, Olav Hohmeyer and Rainer Friedrich in Germany, Marcel Boiteux and Ari Rabl in France, Alan Krupnick and Russell Lee in the USA[74]. Different

[73] The results of this research activity have been published in the ExternE volumes published by the European Commission in 1995, 1999 and 2005.

[74] O. HOHMEYER, R.L. OTTINGER, *Social costs of energy*, *op. cit.*; D. PEARCE, *Cost-benefit analysis and environmental policy*, Oxford Review of Economic Policy, Vol. 14, No. 4, winter 1998; D. PEARCE, A. MARKANDYA, E. BARBIER, *Blueprint for a green economy*, Earthscan Publications, London, 1989; D. PEARCE, A. MARKANDYA, *The benefits of environmental policy*, OECD, Paris, 1989; M. BOITEUX, L. BAUMSTARK, *Transports: choix des investissements et coûts des nuisances*, La Documentation française, Paris, 2001; D. ROSSETTI di VALDALBERO, P. VALETTE, *Environmental and health costs in the European Union: Research and policy*, Encyclopedia of environmental health, Elsevier, 2009.

are the *cost-efficient* analyses that have a focused perspective and reflect the direct financial costs among various options in order to achieve a certain result.

The Jacques Delors White Paper on growth, competitiveness and employment published in 1993 provided the European institutional framework to the concept of external costs[75].

In its Chapter 10 entitled "Thoughts on a new development model for the Community", one can read:

> The under-use of labour is combined with an over-use of environmental and natural resources (...). Because market prices do not reflect sufficiently the limited availability of those natural resources and the environmental scarcities related to their consumption, their over-use has become systematic. This situation cannot continue for different reasons (...). These inefficiencies represent significant but hidden welfare losses. As current economic accounting does not reflect unpriced resources such as the environment, only partial estimates are available. It is, for example, generally recognised that the external cost of current transportation systems alone amount to at least 3%-4% of GDP (environmental pollution, accidents, traffic jams). The more research intensifies and the more is known about these hidden costs which someone ultimately has to bear, the more those estimates become a cause of concern. Another inefficiency of the current development model concerns the external effects related to the use of fossil and nuclear energy. Although intensified technological progress is able to solve many problems, it is also true that energy can no longer be seen as an unlimited resource, particularly if the external costs associated with climate change, with acidification, with health risks, and with nuclear waste and risk are not taken into account. The relative position of energy in the new development model is therefore a key element to be considered. A closer look at the way the Community uses its labour and environmental resources highlights some fundamental weaknesses in the incentive structure of the Community economy attributable to public intervention (e.g. tax treatment of labour costs, transport infrastructure) as well as to market forces (environmental externalities)[76]. As a consequence, it is open to question whether, to an increasing extent, the economic growth figures do not reflect illusionary instead of real economic progress and whether many traditional economic concepts (e.g. GDP as traditionally conceived) may be losing their relevance to policy formulation in the future.

---

[75] European Commission, *White Paper on growth, competitiveness and employment: the challenges and ways forward into the 21st century*, COM(93)700.

[76] These statements were largely coming from the first European research studies on the "double dividend", i.e. positive for the environment and for the employment made with the MIDAS and HERMES models. V. DETEMMERMAN, E. DONNI, P. ZAGAME, *Increase of taxes on energy as a way to reduce CO_2 emissions: problems and accompanying measures*, Report for the European Communities, DG XII, July 1991.

Two years later, the White Paper on an energy policy for the EU was more detailed on the internalisation of external costs[77]. It stated (own translation from French):

> For the energy sector, this could increase the relative price compensated by a parallel modification of the labour charges (...). Members States could consider if some receipts could be reinvested to stimulate the improvements of energy efficiency and the investments in cleaner technologies.

In 2000, the Green Paper: towards a European Strategy for the security of energy supply was published[78]. It mentioned that, "Fiscal instrument (...) should lead to the internalisation of damage caused to the environment". In 2002, the EXTERNE synthetic results were quoted (see Table VI.11) in the Commission Staff Working Paper "Inventory of public aid granted to different energy sources"[79].

In 2001, the sixth Environment Action Programme stated that, "Protecting the environment is not only that people aspire to living in a clean and healthy environment but we must also recognise that the costs and other damages caused by pollution and climate change are considerable"[80]. Furthermore, it insisted, "to ensure that those who cause injury to human health or cause damage to the environment are held responsible for their actions".

In the same year, the White Paper on European transport policy mentioned that the EU should be heading "Towards modal rebalance and greater internalisation of external costs which may be achieved by replacing existing transport system taxes with more effective instruments for integrating infrastructure and external costs"[81]. In 2003, the Commission presented a Proposal for a Directive on the charging of heavy goods vehicles for the use of certain infrastructures. A clear reference to external costs was made:

---

77 European Commission, *An energy policy for the European Union*, COM(95)682. In 1997, the International Energy Agency raises the role of government in the process of technology deployment. Beyond market barriers, market failures are at the origin of an inefficient use of resources. Governments should (only) tackle these market failures, for example, by obliging firms to internalise their external costs. IEA, *Enhancing the market deployment of energy technology: a survey of eight technologies, op. cit.*

78 European Commission, *Towards a European strategy for the security of energy supply, op. cit.*

79 European Commission, Staff working document, *Inventory of public aid granted to different energy sources*, SEC (2002)1275.

80 European Commission, Sixth Environment Action Programme, *Environment 2010: our future, our choice*, COM(2001)31.

81 European Commission, *White Paper on European transport policy for 2010*, COM(2001)370.

> With regard to accident costs (...) a number of studies, past or in progress, use methods that enable monetary values to be attributed to externalities (...).Congestion in certain regions of Europe and along certain major arteries increased in the 1990s and this is now undermining economic competitiveness. According to the most recent study on the subject, external congestion costs due to road traffic alone represent about 0.5% of Community gross domestic product (GDP) and could reach 1% of GDP in 2010, or approximately € 80 Billion[82].

The main reference study quoted was the RECORDIT project ("Real cost reduction of door-to-door intermodal transport") funded by the fifth RTD Framework Programme (Transport programme) and coordinated by the Italian Instituto di Studi per l'Informatica e i Sistemi (ISIS – Andrea Ricci). RECORDIT followed the EXTERNE methodology.

A Directive on the charging of heavy good vehicles for the use of certain infrastructures was adopted in 2006 and it mentions that

> to enable an informed and objective decision to be taken in the future regarding the possible application of the 'polluters pays' principle for all modes of transport, by means of the internalisation of external costs, uniform calculation principles should be developed, based on scientifically recognised data (...). The Commission should begin work on developing a generally applicable, transparent and comprehensible model for the assessment of external costs for all modes of transport to serve as the basis for future calculations of infrastructure charges (...). No later than 10 June 2008, the Commission shall present, after examining all options including environment, noise, congestion and health-related costs, a generally applicable, transparent and comprehensible model for the assessment of all external costs to serve as the basis for future calculations of infrastructure charges. This model shall be accompanied by an impact analysis of the internalisation of external costs for all modes of transport and a strategy for a stepwise implementation of the model for all modes of transport[83].

At the beginning of 2008, the European Commission released a Handbook with estimates of external costs in the transport sector. This Handbook is largely inspired from EXTERNE that is abundantly quoted (Bottom-up Impact Pathway Approach, air pollution costs, figures for life years lost, etc.)[84].

---

82 European Commission, *Proposal for a Directive amending Directive 99/62/EC ("Eurovignette") on the charging of heavy goods vehicles for the use of certain infrastructures*, COM(2003)448.

83 European Parliament and Council *Directive 2006/38 amending Directive 1999/62/EC on the charging of heavy good vehicles for the use of certain infrastructures*, OJ, L 157, 9/6/2006.

84 This study on "Internalisation measures and policies for all external costs of transport (IMPACT)" has been performed by CE Delft (H.P. van Essen, B.H. Boon, R. Smokers, and A. Schroten), INFRAS (M. Maibach, C. Schreyer and D. Sutter),

The 2004 Environmental liability Directive aims at establishing a framework to prevent "significant environmental damage" or to rectify damage after it has occurred[85]. It forces industrial polluters ("operators") to pay prevention and remediation costs of these environmental damages that include biodiversity (whether protected at EU or national levels), waters (covered by the Water framework Directive) and human health (including land contamination when it is a threat to human health). *De facto*, operators carrying out "hazardous" activities are held strictly liable (i.e. no need to demonstrate fault or negligence) for preventing or restoring any damage caused by those activities to land, water and protected habitats and species. Member States may exempt operators from clean-up costs if the damages are caused by pollution released within the terms of emission permits or occurred despite the use of best practice. The damages from nuclear and maritime accidents fall outside the regime's scope and remains subject to existing treaties.

The EXTERNE methodology was used in several EU policies, especially in the cost-benefit analysis. For example, many cost-benefit analysis in the 1990s used the € 3 Million for the average "Value of Statistical Life" (VSL) coming from the EXTERNE publications[86]. In a new research project called NEWEXT ("New elements for the assessment of external costs from energy technologies") funded by the fifth RTD Framework Programme, from empirical surveys on the willingness to pay to avoid a certain damage in several countries of the EU, the new VSL was estimated € 1 Million[87]. From 2000, this new value will be used. One example is the draft Directive on the charging of heavy goods vehicles where the cost for a fatal accident is estimated at € 1 Million per case[88].

Fraunhofer-ISI (C. Doll) and the University of Gdansk (B. Pawlowska and M. Bak). European Commission study (DG TREN), *Handbook on estimation of external costs in the transport sector*, Delft, 2008.

85 European Parliament and Council Directive 2004/35/EC on environmental liability with regard to the prevention and remedying of environmental damage, OJ, L 143, 21/04/2004.

86 European Commission, *ExternE: Externalities of Energy, Methodology 1998 update*, Vol. 7, OPEC, Luxembourg, 1999. The value of statistical life estimates the individual willingness to pay for a marginal change in the risk of death as a consequence of an increase of air pollution. The value of statistical life can be fractioned into the value of a year lost.

87 European Commission, *ExternE: Externalities of Energy, Methodology 2005 update*, OPEC, Luxembourg, 2006.

88 European Commission, *Proposal for a Directive amending Directive 99/62/EC ("Eurovignette") on the charging of heavy goods vehicles for the use of certain infrastructures, op. cit.*

Considering the damages provoked mostly form air pollution on human health, the external costs of electricity and transport could achieve around € 120 Billion per year in the EU[89].

EXTERNE was mentioned and used in various EU energy and transport policies, its main applications are, nevertheless, in the environmental field. Most probably, it is also because "Environment" is the only policy that explicitly requires from the Treaty of the European Union (article 174) that "In preparing its policy on the environment, the Community shall take account of available scientific and technical data (...) and the potential benefits and costs of action or lack of action"[90]. Concretely, EXTERNE provided important inputs to the economic evaluations of the EU strategy to combat acidification and to the large combustion plant Directive[91].

For more than ten years, the whole EU legislation on air quality has been directly and significantly influenced by the European research activities on external costs. The influence starts with the Directive on ambient air quality assessment and management ("Framework Directive") and extends to the first, second and third "Daughter Directives"[92]. In particular, the main economic evaluation of the benefits of the Community strategy to combat acidification, i.e. to limit the emissions of SO_2, NO_x and dust was coming from EXTERNE[93]. The Communication stated:

---

89 D. ROSSETTI di VALDALBERO, *Quantification and tentative internalisation of energy external costs in the EU*, in C. SPITZER, U. SCHMOCKER, V.N. DANG (eds.), *Probabilistic Safety Assessment and Management*, Vol. 4, Springer, London, 2004.

90 European Union, *Consolidated versions of the Treaty on European Union and of the Treaty establishing the European Community*, OJ, C 321, 29/12/2006.

91 As background in this field, the following legal texts have to be considered: Council *Directive 88/609/EEC on the limitation of certain pollutants into the air from large combustion plants*; Council *Directive 93/12/EEC relating to the sulphur content of certain liquid fuels*; Council *Directive 96/61/EC on integrated pollution prevention and control*; and legislation relating to control of emissions from mobile sources (passenger cars and heavy duty vehicles).

92 Council *Directive 96/62/EC on ambient air quality assessment and management*, OJ, L 296, 21/11/1996 ("Framework Directive"); Council *Directive 1999/30/EC relating to limit values for sulphur dioxide, nitrogen dioxide and oxides of nitrogen, particulate matter and lead in ambient air*, OJ, L 163, 29/6/1999 ("First Daughter Directive"); European Parliament and Council *Directive 2000/69/EC relating to limit values for benzene and carbon monoxide in ambient air*, OJ, L 313, 13/12/2000 ("Second Daughter Directive"); and European Parliament and Council *Directive 2002/3/EC of the relating to ozone in ambient air*, OJ, L 67, 9/3/2002 ("Third Daughter Directive").

93 European Commission, *Community strategy to combat acidification*, COM(97)88; and Council *Directive 1999/30/EC relating to limit values for sulphur dioxide, nitro-*

> The dominating economic benefits were to be found in reduced damage to human health, primarily as a result of reduced levels of the secondary pollutants sulphate and nitrate aerosols, and ozone (…). In summary, the incremental annual benefits of the additional emission reductions were estimated to amount to nearly ECU 20 Billion in 2010.

Later in the Communication, another reference to EXTERNE was mentioned:

> The economic cost of the damage resulting from 1 tonne of SO_2 emissions in the Community is approximately ECU 4000: the majority (80%) of these costs being attributed to damage to human health[94].

In 2001, the Clean Air for Europe Programme (CAFÉ) was presented by the Commission[95]. The Directive on national emission ceilings for certain atmospheric pollutants was adopted at the end of the year[96]. The Directive provided the national emission ceilings for SO_2 (sulphur dioxide), NOx (nitrogen oxides), VOC (volatile organic compounds) and NH_3 (ammonia), to be attained by 2010 by the EU Member States. For example, by 2010, the NOx emission ceilings have been fixed at 810,000 tonnes for France and 1,051,000 tonnes for Germany. The VOC (emission ceiling) have been fixed at 159,000 tonnes in Austria and 185,000 tonnes in The Netherlands. In total, for the EU-15, the emission ceilings have been fixed at 3,634,000 tonnes for SO_2, at 5,923,000 tonnes for NOx and at 5,581,000 for VOC by 2010.

EXTERNE was also used to underline the benefits from reducing the emissions from boilers and gas turbines over the period 2000-2010 in the draft Directive to limit emissions of certain pollutants into the air from large combustion plants[97]. The Communication stated:

> The monetary valuation has been carried out in the context of the EXTERNE projects, taking into account only the damage to human health,

gen dioxide and oxides of nitrogen, particulate matter and lead in ambient air, op. cit.

94 European Commission, *Community strategy to combat acidification, op. cit.*

95 European Commission, *Clean Air For Europe (CAFE)*, COM(2001)245.

96 European Parliament and Council *Directive 2001/81/EC on national emission ceilings for certain atmospheric pollutants*, OJ, L 309, 27/11/2001.

97 European Commission, *Proposal for a Directive amending Directive 88/609/EEC on the limitation of emissions of certain pollutants into the air from large combustion plants*, COM(98)415; and European Parliament and Council *Directive 2001/80/EC on the limitation of emissions of certain pollutants into the air from large combustion plants*, OJ, L 309, 27/11/2001.

materials and crops (...). Approximately 99% of the quantified benefits are related to human health effects – of which the largest part is mortality[98].

In total, the benefits are estimated at more than ECU 38 Billion for ten years against a cost of the proposed (amendment) Directive estimated at ECU 2 Billion for ten years.

The 2004 Commission Environmental Technologies Action Plan stated that:

> Getting prices right requires the systematic internalisation of costs through market-based instruments (e.g. taxes, tax breaks, subsidies, tradable permits and deposit – refund schemes). If properly applied they are the best way to reduce pollution because they make producers and consumers bear the real costs of their actions or change their behaviour in a cost-effective way. Moreover, price distortions have been identified as a significant barrier to environmental technologies during the development of this Action Plan. Failure to remove them would significantly reduce the effectiveness of the actions proposed and thus the Action Plan's overall impact. In addition, market-based instruments can also boost the markets for environmental services and increase the demand for public goods[99].

Table VI.9: Scientists results (EXTERNE) used in the 1998 Commission Communication on limitation of certain pollutants into the air from large combustion plants (quantified benefits from reducing emissions from boilers and gas turbines over the period 2000-2010)

Benefit category	Boilers (ECU Million)	Gas turbines (ECU Million)
Materials	40	150
Acute morbidity	270	1,400
Chronic morbidity	15	70
Acute mortality	1,300	6,500
Chronic mortality	4,700	24,000
Total	6,324	32,120

In the 2005 thematic strategy on air pollution, the cost-benefit analysis was essentially based on the EXTERNE work as indicated clearly in the Communication and the Impact Assessment.

> The analysis underpinning this Strategy builds upon Community research on air pollution, including the health impacts of particles, performed under the various RTD Framework Programmes. The Strategy also builds on economic-environmental models and tools developed using Community RTD

98 European Commission, *Proposal for a Directive amending Directive 88/609/EEC on the limitation of emissions of certain pollutants into the air from large combustion plants*, *op. cit.*

99 European Commission, *Stimulating technologies for sustainable development: An environmental technologies action plan for the European Union*, COM(2004)38.

funds (Health valuation from EXTERNE and NEWEXT projects; macro-economic analysis used the general equilibrium model GEM-E3)[100].

The method used for the strategy was firstly to establish a Baseline showing air pollution up to 2020 if no extra measures or additional legislation are implemented. This was then set against Community long-term objectives of achieving levels of air quality that "do not give rise to significant negative impacts on and risks to human health and the environment"[101]. This includes no exceedence of critical loads and levels for natural ecosystems. Then, various scenarios were examined to close the "gap" between the Baseline and the achievement of the long terms objectives. Based on the cost-benefit analysis, interim objectives for the Strategy were set. The cost-benefit analysis was done through a study performed by the UK AEA Technology (Mike Holland and Paul Watkiss), a key partner in the previous EU research projects related to EXTERNE and using this methodology.

Table VI.10: Scientists results (EXTERNE-based Impact Assessment) used in the 2005 Commission strategy on air pollution (human health benefits of the strategy in 2020)

Ambition level	Life Years Lost (million) Fine particulate matter only	Premature deaths (thousands) Particulate matter and ozone	Range in monetised health benefits per annum (€ Billion)
2000	3.6	370	
Baseline 2020	2.47	293	
Strategy 2020	1.91	230	42-135

The EXTERNE results were also used in the European State aid legislation and in particular in the Community guidelines on State aid for environmental protection[102]. Compared to an electricity generation cost of about € 0.04 per kWh, EXTERNE showed that to internalise the external costs would significantly increase (almost double) the costs of producing electricity from coal or oil and moderately (by about 30%) increase the costs of producing electricity from gas.

---

100 European Commission, *Thematic strategy on air pollution*, COM(2005)446; European Commission, *Staff working paper on the Thematic strategy on air pollution*, SEC(2005)1132; European Commission, *Staff working paper (Impact Assessment) on the Thematic strategy on air pollution and the Directive on ambient air quality and cleaner air for Europe*, SEC(2005)1133.

101 European Commission, *Sixth Environment Action Programme – Environment 2010: our future, our choice, op. cit.*

102 European Commission, *Community guidelines on state aid for environmental protection*, OJ, C 37, 3/2/2001.

From the synthetic Table VI.11) published in the EXTERNE – National Implementation (case studies in several Member States), one can see that coal and lignite produced electricity provokes external costs of a range included between € 0.02-0.15 per kWh and that natural gas produced electricity has external costs included in a range included between € 0.01-0.04 per kWh. Hydro and wind produced electricity have external costs between € 0.0005-0.01 per kWh[103]. *De facto*, a realistic average of damages coming from fossil fuels *versus* renewable energy sources could be considered at € 0.05 per kWh.

The Community guidelines on State aid for environmental protection mentioned that:

> the principle of "prices to reflect costs" states that the prices of goods or services should incorporate the external costs associated with the negative impact on the environment of their production and marketing (…). Member States may grant operating aid to new plants producing renewable energy that will be calculated based on the external costs avoided (…). At any event, the amount of the aid thus granted to the renewable energy producer must not exceed € 0.05 per kWh[104].

In the transport sector, the Commission (DG Energy and Transport) has accepted several times that Member States gave subventions (the Commission raises no objection to the State Aid) to rail or waterway transport that compensate for the avoided external costs from the road transport.

This occurs, for example, for rail in an Italian State Aid case (Bolzano-Brenner journey) or for waterway transport in an Austrian and Belgium State Aid cases[105]. In the three cases, the Commission grounded its decision on scientific studies that compared the external costs of freight transport by road and by train or waterway[106].

In the three cases, a monetary value was given. For instance, in the first case, the Commission authorised the Italian Government to give a maximum subvention to € 33 for the Bolzano-Brenner journey due to the avoided external costs that include accidents, noise, pollutants, CO_2 and other ecosystem damages. In the Austrian and Belgian cases, the

---

103 European Commission, *ExternE: Externalities of Energy, National Implementation*, Vol. 10, OPEC, Luxembourg, 1999.

104 European Commission, *Community guidelines on state aid for environmental protection*, *op. cit.*

105 State Aid n° 638/2001 – Italy; State Aid n° 219/2001 – Austria; State Aid n° 4/2004 – Belgium.

106 B. DURANTE, *Analisi dei costi esterni non corrisposti da chi effettua il trasporto nel corridoio Bolzano-Brennero*, Bolzano, 2001; Herry/Trafico, Externe kosten in Güterverkehr in Österreich, 2000.

Commission has approved a State Aid for the avoided "external costs" (between inland navigation and road) on the river Danube and on the Walloon inland waterway that could be up to € 27 for a 45 foot intermodal loading unit.

Table VI.11: Scientists results (EXTERNE) used in the 2001 Community guidelines on state aid for environmental protection (sub-total of quantifiable externalities such as global warming, public health, occupational health, material damages)[107]

	Coal and lignite (€ cents/kWh)	Natural Gas (€ cents/kWh)	Hydro (€ cents/kWh)	Wind (€ cents/kWh)
Austria	-	1-3	0.1	-
Belgium	4-15	1-2	-	-
Denmark	4-7	2-3	-	0.1
Finland	2-4	-	1	-
France	7-10	2-4	1	-
Germany	3-6	1-2	-	0.05
Greece	5-8	1	-	0.25
Ireland	6-8	-	-	-
Italy	-	2-3	0.3	-
Luxembourg	-	-	-	-
Netherlands	3-4	1-2	-	-
Portugal	4-7	1-2	0.03	-
Spain	5-8	1-2	-	0.2
Sweden	2-4	-	0-0.7	-
UK	4-7	1-2	-	0.15

107 For clarity and because other debates surrounded the figures on nuclear, on biomass and on photovoltaic, these figures have been taken out from the table.

CHAPTER VII

Models *ex-post* Evaluation and Comparison

Summary

Various economic and energy modelling activities (Top-down and Bottom-up) are pursued by public authorities, private companies and research institutes with the aim to provide energy forecasts and to assess the impact of energy-related policies (See Chapters II, III, IV and V).

The five European case studies – on energy taxation, on climate change policy, on promotion of renewables energy sources, on energy efficiency, and on the measurement and internalisation of external costs – dealt mainly with *ex-ante* evaluation, i.e. models were used to assess the impact of a specific policy before to be taken (See Chapter VI). Only a very few *ex-post* evaluations of modelling activities have been carried out at the EU-level.

The first part of this Chapter investigates the assumptions and the results from a European study carried out in the middle of the 1980s with the targeted year of 2000 as presented in the European Commission "ENERGY 2000" report. Concretely, modelling results (forecasts) are compared with real statistical data[1].

ENERGY 2000 is based on the combination of outputs coming from the energy demand-forecasting model MEDEE and the supply techno-economic energy model EFOM developed since the 1970s. "ENERGY 2000" published in 1986 is the main European modelling study in the energy field for which real data are now available.

Three kinds of *ex-post* evaluations could have been done: by final sector (industry, transport, residential-tertiary), by country (each of the 10 Member States analysed in ENERGY 2000) or by fuel and technology (solid fuels, oil, natural gas, nuclear and renewables). This last one has been preferred for reasons of availability of precise public data and for not entering in specific national (or sometimes even regional) policies.

1 Part of this chapter has been published in P.A. PILAVACHI, Th. DALAMAGA, D. ROSSETTI di VALDALBERO, J.-F. GUILMOT, *Ex-post evaluation of European energy models*, *op. cit.*

If the models projected the total energy consumption for the ten European Community Member States at a 15 years time horizon in an excellent way (less than 2% deviation between real statistics and model results), relatively large differences exist between statistics and projections concerning individual fuel and technologies. For example, the models sub-estimated by 30% the natural gas consumption in the European Community of that time.

Rather than weaknesses of the model assumptions and structure (population and economic growth, costs of the technologies, etc.), these large differences are essentially due to unanticipated strong political decisions and social behavioural changes that were not part of the reference scenario such as the:

- EU legislation to protect the environment at the end of the 1980s (cf. Directive on the limitation of emissions of pollutants in large combustion plants) and at the international level in the 1990s (Rio Summit and Kyoto Protocol);
- The closing of mines in the UK or in France (see the strong model overestimation for solid fuels);
- New market mechanisms elaborated essentially during the 1990s to introduce more competition in energy markets – liberalisation of electricity and gas markets;
- Abrogation of the gas directive that prohibited natural gas to be used for base-load power generation (see the strong model underestimation of natural gas);
- Strong support for the development of renewable energy sources like the Feed-in tariff laws in some countries;
- Large increase in the transport demand in the 10 EC Members States probably also connected with the air transport liberalisation during the projection period and to the improved easiness to move within Europe (see "Single European Act" signed in 1986).

It is worthwhile mentioning, especially regarding statistics on renewables, that the availability of data on these energy sources and technologies was very poor at the beginning of the 1980s and that some definitions have changed over the projection period.

The second part of this Chapter aims at "validating" or "benchmarking" the model results by a comparative method, i.e. comparing three modelling-based studies published in 2006 by the European Commission (WETO using POLES), the International Energy Agency (WEO using WEM) and the US Department of Energy (IEO using SAGE)[2].

---

2 Rather than a comparison, another possible method to "validate" or to check the preciseness of a model is to run it at the reverse (*faire tourner le modèle à l'envers*).

For the main exogenous hypothesis, i.e. demographic dynamics (world population) and economic growth (GDP), similar assumptions are made in the three studies although the US-IEO study assumes a higher increase of GDP particularly for the decade 2020 to 2030.

The comparison of the main results of the "Reference case" of the three studies demonstrates good agreement among studies concerning the global energy consumption: EC-WETO estimates 17,000 Mtoe by 2030 and this is in the middle-range of the two other studies. For CO_2 emissions the IEA-WEO study estimates around 40 Gt CO2 in that is in the middle-range of the two other studies.

Important differences in the projections appear for individual energy sources and technologies by 2030. Whereas the prospects for oil by 2030 are very similar (around 33% of the total energy consumption), those for coal and gas exhibit more noticeable differences, mainly driven by the fuel mix in the power sector.

The US-IEO forecasts significantly higher coal consumption (27%) than the other studies. In the EC-WETO study, the lower consumption of coal (24%) is partially offset mainly with a higher consumption of natural gas in the decade 2010 to 2020, and nuclear energy in the decade 2020 to 2030. Consequently, the EC-WETO turns out to be the reference scenario with the highest natural gas consumption in 2010 and 2020 (falling back to an average level in 2030) and with the most pronounced nuclear electricity production in 2030 (more than 15% against half of it in the IEA-WEO study).

The forecasts for hydro electricity and biomass are relatively similar for EC-WETO and IEA-WEO, with the latter showing slightly higher values for 2030. Although high growth rates for other renewables (wind and solar technologies) are foreseen in all scenarios, the contribution of other renewables to overall world energy consumption remains at a around 5%, even in the most favourable case (EC-WETO). For the share of renewable sources in total electricity generation, the three studies are well in line with about 20% of green electricity by 2030.

"Afin de bien vérifier la précision d'un modèle, indépendamment de l'aléa qui porte sur les variables exogènes, un exercice consiste à réaliser des simulations sur le passé en donnant aux variables exogènes leurs vraies valeurs. On peut alors à l'aide de statistiques précises apprécier l'aptitude du modèle à décrire la réalité économique". P. ZAGAME, *L'expérience française de modélisation macro-économétrique: bilan et perspectives*, *op. cit.* Part of this chapter is largely based on a European Commission Working Paper (JRC-IPTS): SCHADE B., WIESENTHAL T., *Comparison of long-term world energy studies – assumptions and results from four world energy models*, JRC-IPTS, Sevilla, March 2007.

EU "ENERGY 2000" study with MEDEE and EFOM

Since 1975, before the existence of the EU Research Framework Programmes (See Chapter II), the first European Community non nuclear energy programme provided support to the development of energy models ("Energy Systems Analysis" programme)[3].

The MEDEE model (*Modèle de prospective de la Demande Energétique à long terme*) and the EFOM model (*Energy Flow Optimisation Model*) were originally developed in the 1970s by the Energy Economic Institute of the University of Grenoble (IEJE transformed later in IEPE and today called LEPII/EPE). Nowadays, MEDEE is used by the French energy consultancy firm ENERDATA.

As indicated by Bertrand Château and Bruno Lapillonne at the end of the 1970s, MEDEE is a long-term energy demand foresight model, which forecasts energy demand in the European Community of that time. It employs mathematical algorithms and techno-economic data. MEDEE:

- Disaggregates the total energy demand into relevant homogeneous categories;
- Systematically analyses the social, economic and technological factors determining the evolution of the long term energy consumption;
- Organizes all factors into a hierarchical structure;
- Constructs an associated energy scenario.

In the MEDEE approach, energy demand is induced by socio-economic determinants that is, by economic activities and by the satisfaction of social needs (e.g. mobility of persons and temperature in rooms). Bertrand Château and Bruno Lapillonne pointed out that these determinants lead to a demand of useful energy (e.g. process heat and mechanical energy) whose intensity depends on the technologies used to satisfy social needs or to perform the economic activities. The demand for energy commodities or final energy (e.g. coal, electricity and gas) can be calculated from the level of useful energy demand which will depend upon the efficiency of the equipment (e.g. furnaces, boilers and engines) used to convert the final energy into useful energy. Thus, the final energy demand of a society is directly related to its social, economic, and technological pattern of development.[4]

---

3 R. RUBIN, E. ROMBERG, *Energy models for the European Community*, in A. STRUB (ed.), *Energy models for the European Community*, *op. cit.*

4 B. LAPILLONNE, B. CHATEAU, *The MEDEE models for long-term energy demand forecasting*, International Conference of Dublin, Energy systems analysis,

MEDEE contains the residential, tertiary, transport and industrial sectors. The model application is related with the population and the economic growth of each European country. The model reflects the general behaviour of the population with regard to energy use. It also considers the energy policies applied by national governments[5].

As an example, for the residential sector (a similar approach was applied to the other sectors), the model:

- Takes into account assumptions related to demography (population, persons per household) and the evolution of the structure of the population.
- Takes into account exogenous assumptions related to housing and governmental social policy. It calculates the number of dwellings withdrawn from the stock, the number of dwellings built to replace the demolished dwellings and to face the increase in number of households.
- Calculates the heat requirement expressed in useful energy and takes into account the insulation standards and household behaviour such as indoor temperature.
- Simulates the choice of heating appliances in new dwellings.
- Calculates the evolution of the heating systems split into different types of dwellings, accounting for policy and technical constraints due to global and sector energy policies.
- Calculates the final energy demand, by energy product and globally.

The EFOM model complemented the MEDEE model. As mentioned by Dominique Finon in 1979, EFOM is a linear programming model aimed to represent the energy transformation system (electricity production, refineries, etc.) and the global energy supply (primary energy production, imports and exports). It is a supply techno-economic energy model that simulates and/or optimizes the primary energy requirements and the related investments in energy production and consumption necessary to satisfy a given exogenous demand of useful or final energy. The main features of this energy model were that techno-economic information was stored in a database designed to allow the user to build particular energy structures from the very large national energy systems; there was the possibility for simulation and/or optimization.

---

op. cit., M. CAMOS, A. DUMOAT, P. VALETTE, *MEDEE 3 – Modèle de demande en énergie pour l'Europe*, EC-TEC-DOC, Paris-Luxembourg, 1986.

5 B. CHATEAU, B. LAPILLONNE, "Long-term energy demand simulation", in A. STRUB (ed.), *Energy models for the European Community*, *op. cit.*, B. LAPILLONNE, B. CHATEAU, "The MEDEE model for long term energy demand forecasting", *Socio-Economic Planning Sciences*, Vol. 15, issue 2, 1981.

The energy system was represented by an oriented network in which primary energy was cascading and was gradually transformed in order to satisfy the final energy demand. This final energy was transformed further down to useful energy so as to satisfy an exogenous demand. The network was a concatenation of links with upstream and downstream nodes. Nodes were representatives of intermediate status of energy forms while links represented activities (extraction, power production, transport, distribution and others)[6].

In 1983, three socio-economic scenarios were studied by the European Commission (former DG XVII-Energy, now merged with transport into the DG Transport and Energy) representing three theoretical sets of development for the Community economy up to year 2000:

- A scenario called "Open Competition" which assumed full-scale economic liberalism thus minimising public intervention;
- A scenario named "Cooperation" based on a stimulus of world economic growth as a result of a successful North-South dialogue;
- A third scenario called "Europe" which assumes fall back on itself to both economic and energy problems.

As indicated by Jean-François Guilmot, David Mc Glue, Pierre Valette and Christian Waeterloos

> these rather exaggerated scenarios enabled the outer limits to the Community energy markets in 2000 to be explored and the problem areas and constraints to be identified. But, in view of the rather theoretical nature of this exercise it was judged thereafter to be useful to define a fourth scenario of a more realistic nature as a basis for reflections within the Commission on Community energy policy in the medium and long-term, and specifically as a background to the definition of new Community energy policy objectives[7].

In 1985, MEDEE and EFOM models were used to forecast future energy consumption (for the industrial, residential, transport and tertiary sectors) for the year 2000 in the so-called "ENERGY 2000" study[8]. The European Community of that time was composed of ten Member States (EC-10): Belgium, Denmark, Germany (RFA), Greece, France, Ireland, Italy, Luxembourg, The Netherlands and United Kingdom.

The "ENERGY 2000" reference scenario is a "Business as usual" case based on the assumption of no major changes to the existing eco-

---

6 D. FINON, "Scope and limitations of formalized optimization of a national energy system", in A. STRUB (ed.), *Energy models for the European Community*, *op. cit.*

7 European Commission, *Scenarios to year 2000: a prospective study of the energy market in the European Community*; Internal Working Paper, Brussels, 1983.

8 J.F. GUILMOT, D. McGLUE, P. VALETTE, C. WAETERLOOS, *Energy 2000*, European Commission, Luxembourg, 1986.

nomic systems of the Member States during the period under consideration. The economic framework assumes a gradual resumption of economic growth, considering that the restructuring of industry will lead to a relative reduction in the share of the traditional manufacturing sectors.

The economic framework assumes an average growth rate of Community GDP of 2.6% per annum for the period 1984-2000. This trend will be accompanied by a relatively stable share for private consumption and a reduction of public consumption in favour of investment. The hypothesis adopted concerning the structural development of GDP envisages an increase in the relative contribution of services. Their share was predicted to grow from 55% to 59% from 1980 to 2000. Consequently, there is a decline in the share of industry (-2%), of agriculture (-1%) and of building construction (-1%).

The population of the EC-10 (excluding former Eastern Germany) in 1980 represented about 272 Million inhabitants and a total population of around 280 Million inhabitants was expected by year 2000.

The price of imported crude oil was assumed to go through three successive stages: fall in nominal terms until the 1980s, after the peak of 1980, followed by stable prices in real terms until 1990, and finally progressive growth to reach 35$/bl in 2000 (prices of 1983). It was assumed that the price of oil will remain the "guide price" for energy and that the algorithm for indexing gas prices to crude oil should continue to apply until 2000.

The "ENERGY 2000" study uses scenarios and forecasts the energy consumption in the EC-10. These forecasts were established in 1985 and refer to energy consumptions for 2000. Statistical data and information relative to energy consumption of the EC-10 have been obtained from. Energy consumption data for 2000 of each energy source and technology are compared and results are concluded.

Since the study was produced, the map of Europe has changed considerably with, in particular affecting this comparison, the re-unification of Germany. The "ENERGY 2000" study takes only into consideration Western Germany (BRD) while Eurostat data for year 2000 present the re-unified Germany without identifying the weight of former Eastern Germany (DDR).

Energy policies prevailing in 1983 had to be considered in the construction of this scenario:

- Implementation of large nuclear power investment programmes as well as the construction of several solid fuel power plants under study at the time;
- Application of Directive 75/405/EEC limiting drastically the use of natural gas for power generation;

- Continuation of fiscal policies concerning energy prices; unified at EU level during the 1990s and taxes largely increased since 1983.

Convergence and divergences between statistics and forecasts

Three kinds of comparisons could have been made between the projections coming from the models (MEDEE and EFOM) and the real statistical data (Eurostat):

- By final consumption, i.e. by industrial, transport, and residential/tertiary sectors;
- By country, i.e. each of the 10 Member States analysed in the ENERGY 2000 study;
- By fuel and technology (oil, gas, nuclear and renewables).

This last comparison has been preferred for avoiding to enter in a subject for which the information was not easily publicly available (for example, the sub-division in the transport sector in the first case) and for avoiding in entering too much in the national and sometimes regional policies that largely affect Member States choices.

Figure VII.1: Comparison of total energy consumption between statistics and modelling results in the EC-10 (in Million Toe)

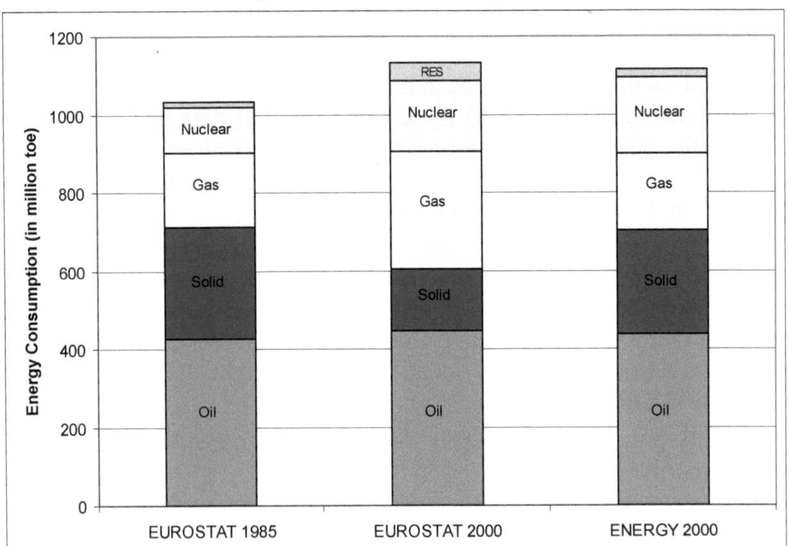

The comparison of total energy consumption in the European Community (EC-10) between Eurostat data (for the year 2000) and the ENERGY 2000 study shows an impressive similarity, i.e. a difference of less than 2% between the real data and the MEDEE and EFOM projections. As shown in Figure VII.1 the difference between statistical data and modelling results is extremely small (considering the subtraction of Eastern Germany from the Eurostat data)[9].

In this European exercise, if the total energy consumption between statistics and forecasts is very close within the 15 years timeframe of the comparison (1985-2000), relatively large differences appear when looking at each energy source (and technology) consumption, i.e. solid fuels (coal, lignite and coke), oil, natural gas and renewables[10].

Observed statistical data and forecasts for the solid fuels in the year 2000, are different essentially due to unexpected political decisions. While the "ENERGY 2000" study forecasted a low decrease, observed data give a substantial reduction in solid fuels consumption by 87 Mtoe.

Figure VII.2: Comparison between statistics and modelling results for solid fuels consumption in the EC-10 (in Million Toe)

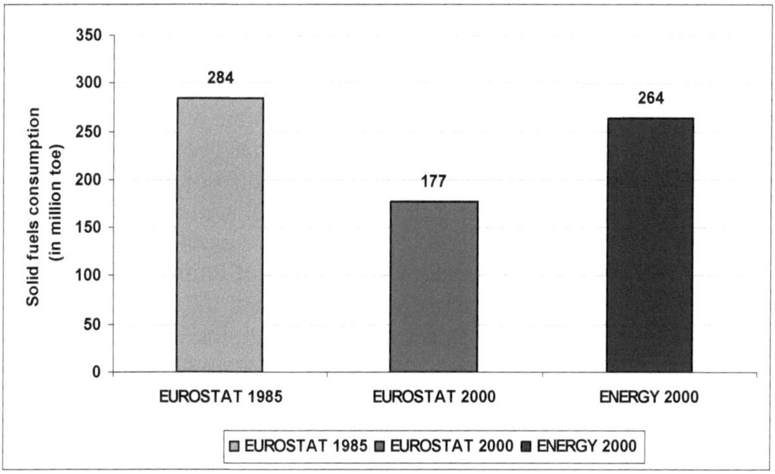

9 The total energy consumption of DDR in 2000 has been estimated at 64 Mtoe considering a population of 16 millions people, a *pro capita* energy consumption of 4 toe and a primary energy consumption shared among oil (34%), solid fuels (30%), natural gas (23%) and nuclear (13%).

10 The following Figures related to EC-10 consumption of solid fuels, oil, natural gas and renewable energy sources do not include DDR.

It has to be noted that the main political factors that have influenced coal consumption in Europe are:

- The reduction of coal subsidies for indigenous European coal production. Consequently, coal production was considerably reduced in most of the Member States.
- Environmental concerns about acid rain that led to Directive 88/609/EEC on the limitation of emissions of certain pollutants into the air from large combustion plants (above 300 MW_{th}). This Directive was very strict for new plants and was often revised in order to be reinforced and to address smaller plants (down to 50 MW_{th}). In Italy for example, the 50% overestimation of coal in ENERGY 2000 (30 Mtoe vs. 15 in reality) is largely due to the massive introduction of natural gas in the power generation sector as a result of various legislations put in place to prevent air pollution while the foreseen development of the Italian power sector at the beginning of the 1980s was mainly based on steam coal.
- The abrogation of the gas directive in 1991 that prohibited natural gas to be used for base-load power generation.
- The restructuring of the iron and steel industry: consumption of coal has been marked by the restructuring of the iron and steel industry reflected by a reduction in capacity but also by the introduction of electric furnaces.
- National policies: Between 1985 and 2000 the production in the two major producers (Germany and the UK) has been reduced by about 50%, while only Greece increased its production level of lignite. In the UK, the very large difference between statistics (35 Mtoe) and projections (70 Mtoe) concerning the consumption of coal is particularly due to three factors: the closing of mining activities in the 1980s by Prime Minister Margaret Thatcher, the use of gas fields in North Sea and the "rush for gas" following the liberalisation of the power sector. In France, coal was largely over-estimated (about 40%) by ENERGY 2000 as few experts were expecting at the beginning of the 1980s, a political decision to close the majority of coal mines by 2000.

Concerning oil, the significant increase in its price, resulting from the crises of 1973 and 1979, led the governments to apply energy policies with main characteristics to replace oil as the energy base, and to strengthen domestic energy supply. As shown in Figure VII.3, the "ENERGY 2000" study forecasted oil consumption to increase slightly during the period 1985 until 2000. This was the result of several assumed opposing trends:

- The steady growth of transport fuel requirements in particular for road and air transport; this latter being accentuated by the air travel liberalisation since the second half of the 1980s;
- An increased requirement for petroleum products for non-energy uses (petrochemicals);
- A slight growth in the consumption of both heavy and light fuel oil by industry up to 1990 as the economy recovered, followed by a fallback in the face of competition from natural gas, steam coal and electricity;
- A steady decline in the use of gas oil for space heating in the residential and tertiary sectors;
- A considerable reduction in the consumption of residual fuel oil in power stations up to 1990, followed by relative stability at 1990 levels until 2000[11].

However, in reality the observed data confirm these trends even if the absolute 2000 observed level is above the forecasted level, even if the inclusion of Eastern Germany consumption in the Eurostat data make the difference smaller.

Figure VII.3: Comparison between statistics and modelling results for oil consumption in the EC-10 (in Million Toe)

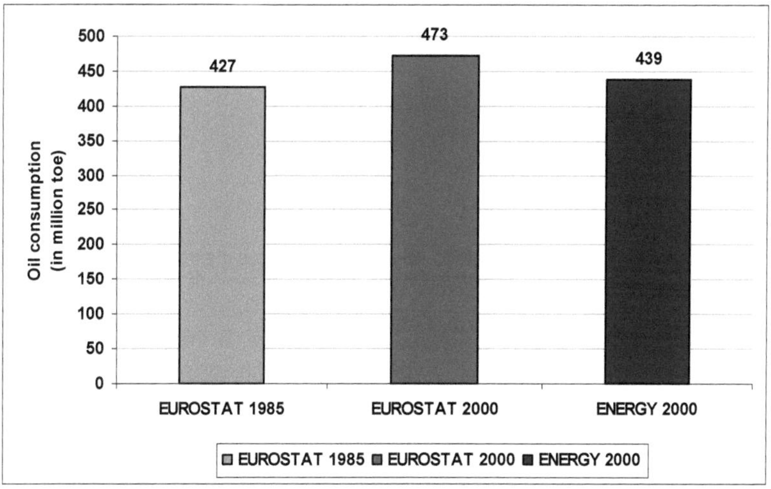

The study "ENERGY 2000" forecasted stability in natural gas consumption. However, as shown in Figure VII.4, the importance of natural gas considerably increased during the 1985-2000 period.

[11] J.F. GUILMOT *et al.*, *Energy 2000*, *op. cit.*

This is due to closing of coalmines in the UK and other Member States, the development of the North Sea natural gas reserves and the political will of countries to reduce greenhouse gas emissions after the 1992 United Nations Rio Summit and the 1997 Kyoto Protocol of the United Nations Framework Convention on Climate Change (UNFCCC).

This period is known as the "Rush for gas" mainly for electricity production where natural gas substituted coal, taking advantage of improved technologies such as Gas Turbine Combined Cycle (GTCC) especially based on natural gas. A clear example is the UK where, in the 1990s, the share of electricity produced from natural gas increased from 2% to 40%[12]. In addition, the process of energy liberalisation in Europe offered a clear advantage to short time investments such as gas plants.

Figure VII.4: Comparison between statistics and modelling results for natural gas consumption in the EC-10 (in Million Toe)

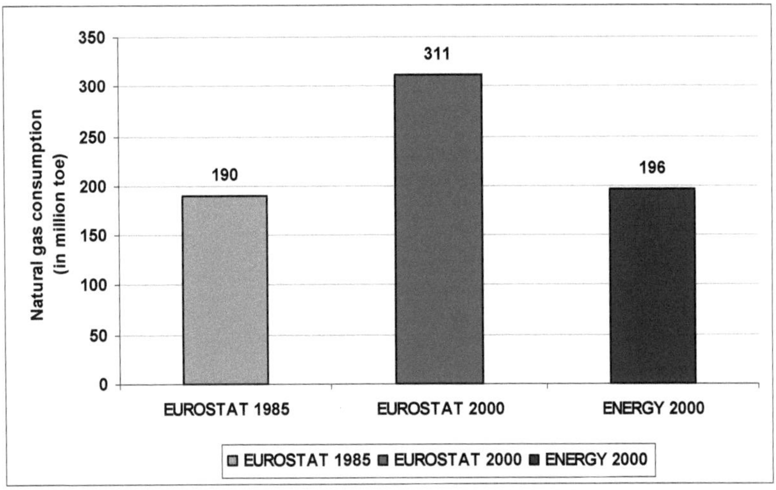

Regarding nuclear energy, as it appears from Figure VII.5, the study successfully projected nuclear energy consumption for year 2000. This accuracy is coming from the delay exceeding 10 years between the decision of investment and the final commissioning of a nuclear unit, due to the long process of authorization and construction time, which is generally greater than 5 years. As the nuclear power stations programme was well known from the authors of the "ENERGY 2000" study, the forecast is accurate for the fifteen-year period.

[12] D. ROSSETTI di VALDALBERO, *L'Europe et le changement climatique: Actualité et prospective*, Revue de l'Énergie, No. 534, Paris, Février 2002.

The study was also elaborated in a period context more favourable to nuclear energy and was not in a position to integrate the slight decrease in use of nuclear energy, which was due to public reaction mainly after the Chernobyl accident that happened on 26 April 1986.

Figure VII.5: Comparison between statistics and modelling results for nuclear energy consumption in the EC-10 (in Million Toe)

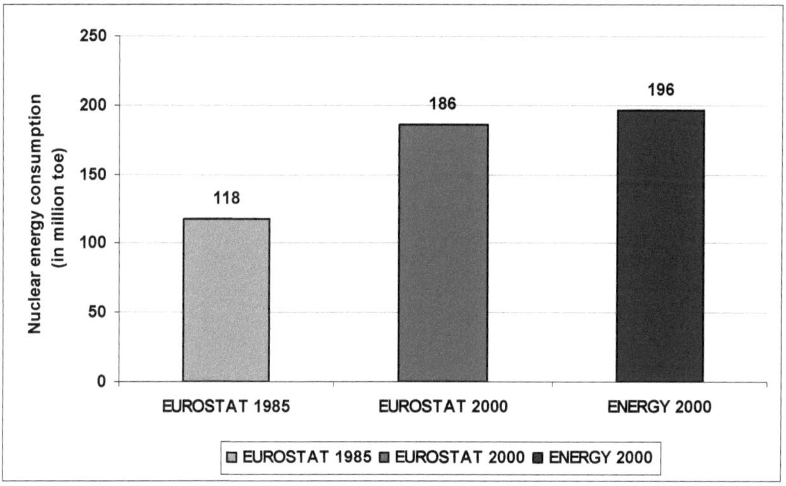

As far as renewable energy sources are concerned, in 1983, when the "ENERGY 2000" study was carried out, the quality of the Eurostat statistical data concerning renewable energy were very poor, with the exception of hydropower and geothermal. Only industrial, domestic and wood waste burned in power stations were taken into consideration. Consequently, the contribution of renewables, as published in the Eurostat energy statistical yearbook in the middle of the 1980s was limited 11 Mtoe of hydropower, 1.8 Mtoe of geothermal energy and only 1.6 Mtoe of other renewables[13]. *De facto*, except large hydro and geothermal, most of the other renewable uses were not accounted for, especially biomass in final sectors i.e. industry and households.

Since then, probably in part because renewables were considered as an important component of European energy policy, substantial efforts were made to establish more complete and accurate renewable statistics. The re-evaluation contribution of renewables for 1985 is today 35.6 Mtoe including 12.8 Mtoe of hydropower and geothermal[14]. Hence,

13 Eurostat – *Energy Statistical Yearbook 1985*, Official Publications of the European Communities, 1987.

14 Eurostat, 2007.

the study predicted an increase from 14.4 Mtoe up to 21 Mtoe, while statistics available now show an increase from 35.6 Mtoe up to 50.6 Mtoe, as shown in Fig.VII.6.

It has also to be stressed that the model did not anticipate the large increase of wind energy especially related to the emergence of proven technologies as well as to specific policy measures supporting the development of green electricity such as the guaranteed "feed-in tariffs" primarily in countries like Denmark, Germany and then Spain[15].

Figure VII.6: Comparison between statistics and modelling results for renewable energy consumption in the EC-10 (in Million Toe)

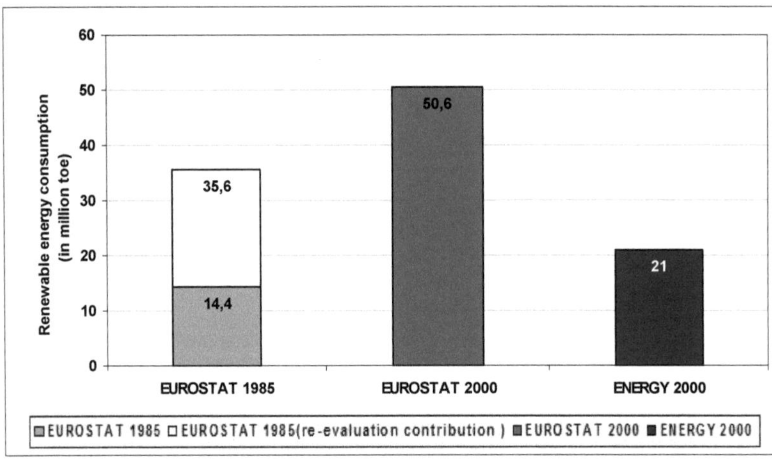

As a conclusion of this part of this Chapter, one can say that although there exist some (large) variations among the modelling projections and the real figures (per energy technology and per country), the overall comparison of total energy consumption in the European Community (EC-10) between Eurostat data (for year 2000) and the ENERGY 2000 study shows an impressive similarity[16]. The difference between statistical data and modelling results is extremely small, i.e. a difference of less than 2% between the real data and the MEDEE and EFOM projections (considering the subtraction of Eastern Germany from the Eurostat data). This is mainly due to the fruitful combination of two kind of modelling tools covering well the energy sector from both

15 *Stromeinspeisungsgesetz* or Electricity Feed-in Law (EFL), Germany, 1990.

16 Only very few ex-post evaluations of the results of modelling activities have been carried out at the European Community level. In the US, a complete exercise on the NEMS model *ex-post* results was accomplished by the Energy Information Administration. EIA, *Annual Energy Outlook – Retrospective Review: Evaluation of Projections in past editions (1982-2006)*, Washington, 2006.

the demand and the supply sides. In addition, the models that allow "to loop the loop" have proved to be effective at the overall energy system level.

EU, international and US energy studies assumptions

In addition to the *ex-post* evaluation of modelling results previously described in this Chapter, another way of "validating" or making a "benchmark" of the modelling exercises is to compare their main assumptions and results to see the concomitancy and discrepancies among different studies[17].

The "Reference case" of three world energy studies (already mentioned in Chapters III and IV) are compared:

- The European Commission WETO report using the POLES model[18];
- The International Energy Agency (IEA) WEO report using the WEM model[19];
- The USA Department of Energy (Energy Information Administration) IEO report using the SAGE model[20].

Only the Reference case projections – and not specific variant scenarios – are compared. As the three studies have a different starting year (2001 for WETO, 2004 for the WEO and 2002 for the IEO), a different time-frame and a different territorial disaggregation (the models used in WETO splits the world in 47 countries/regions, WEO in 21 and IEO in 16), the next tables and figures are provided for the world only and for the simulation years 2010, 2020 and 2030.

With a world population of slightly less than 7 Billion in 2010, around 7.5 Billion in 2020 and around 8 Billion in 2030, the three studies reflect very similar population projections.

---

17 Several model comparative assessments have been done, especially at the European Union, International and American-levels (cf. ACROPOLIS and CASCADE-MINTS European research projects funded under the fifth and sixth Research Framework Programme, the ETSAP implementing agreement of the International Energy Agency and the works of the IIASA and of the EMF. This Chapter is largely based on a European Commission Working Paper (JRC-IPTS): SCHADE B., WIESENTHAL T., *Comparison of long-term world energy studies – assumptions and results from four world energy models*, JRC-IPTS, Sevilla, March 2007.

18 European Commission, *World Energy Technology Outlook – 2050 (WETO-H2)*, Luxembourg, 2006.

19 International Energy Agency, *World Energy Outlook (WEO)*, Paris, 2006.

20 US Department of Energy (EIA), *International Energy Outlook (IEO)*, Washington, 2006.

Figure VII.7: World population assumptions in the EC-WETO, IEA-WEO and US-IEO studies (in Million persons)

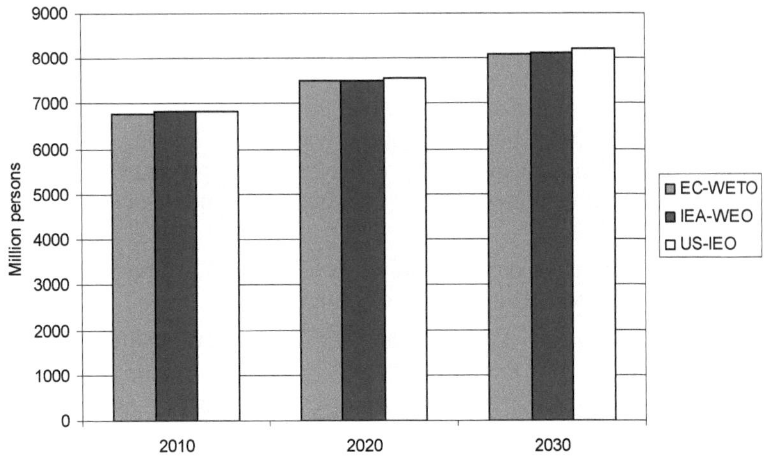

The GDP growth annual rates differ among the three studies with the lowest in the EC-WETO report and the highest in the US-IEO report whether the IEA-WEO is in the middle. From 2010 to 2030, GDP is expected to increase by around 80% in EC-WETO and IEA-WEO, and is expected to more than double in the US-IEO study.

Table VII.1: World GDP annual growth rate assumptions in the EC-WETO, IEA-WEO and US-IEO studies (in percentage)

	Annual GDP growth rate (%)	
	2010-2020	2020-2030
EC-WETO	3.20	2.65
IEA-WEO	3.45	2.90
US-IEO	3.75	3.56

Figure VII.8: World GDP assumptions in the EC-WETO, IEA-WEO and US-IEO studies (in \$ Billion$_{2000}$)

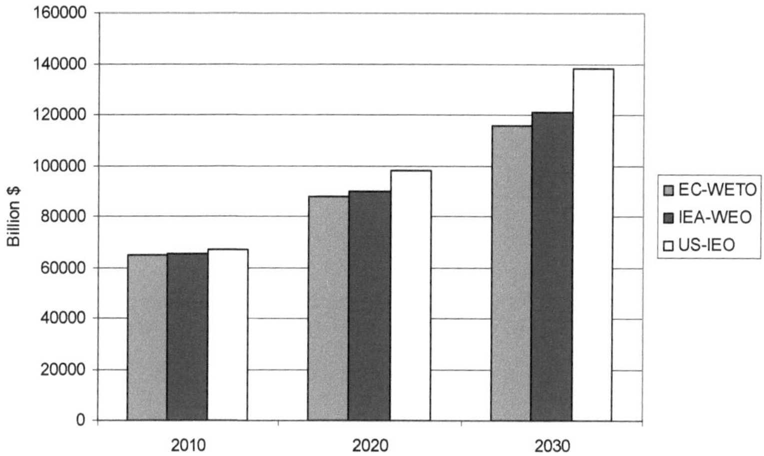

Oil prices remained below 25 \$/bl in the second half of the 1990s but raised to a range of 40-77 \$/bl between 2004 and 2006. These price fluctuations were accompanied by an extremely high volatility mainly due to political uncertainties in several oil producing countries, the impact of weather extremes on production capacities and stockpiling trends[21].

The EC-WETO and the IEA-WEO studies include a mechanism of endogenous price formation in the oil market depending on the reserve-production ratio in the long-term and on the spare production capacity of large oil producing countries in the short term. The US-IEO assumes exogenously the evolution of oil price (at 59 \$/bl for the year 2030) rather than estimating it. In the three studies, the oil price is included between 55 and 70 \$/bl in the year 2030 with the highest price in EC-WETO and the lowest price in the IEA-WEO.

Comparison of modelling-based world energy outlooks

The world energy demand is very similar between the EC-WETO and the IEA-WEO studies at around 12 Gtoe in 2010, 14 Gtoe in 2020 and 16-17 Gtoe in 2030. The US-IEO expects higher energy consumption with 15 Gtoe in 2020 and 18 Gtoe in 2030 that is partly explained by the higher GDP growth assumptions.

21 B. SCHADE, T. WIESENTHAL, *Comparison of long-term world energy studies – assumptions and results from four world energy models*, *op. cit.*

Table VII.2: Projected world energy consumption annual growth rate in the EC-WETO, IEA-WEO and US-IEO studies (in percentage)

	Annual energy consumption growth rate (%)	
	2010-2020	2020-2030
EC-WETO	1.64	1.50
IEA-WEO	1.70	1.31
US-IEO	1.86	1.64

Figure VII.9: Projected world energy consumption in the EC-WETO, IEA-WEO and US-IEO studies (in Mtoe)

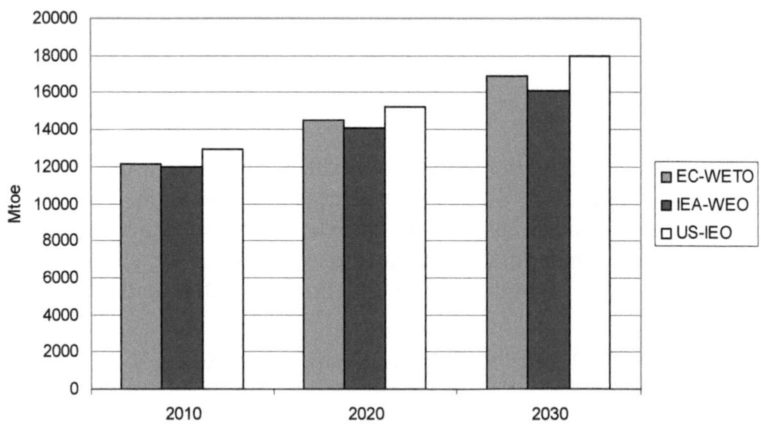

The energy intensity (ratio between primary energy demand and GDP) is expected to decrease by 1.1-1.8%/year in the three studies showing a decoupling between energy demand and economic growth. The reductions in energy intensity reflect the improvements in technical energy efficiency, in consumers' behaviour and in the structural changes of the economy (shift from energy-intensive industries towards services).

Figure VII.10: Projected energy intensity in the EC-WETO, IEA-WEO and US-IEO studies (in Toe per $ Million)

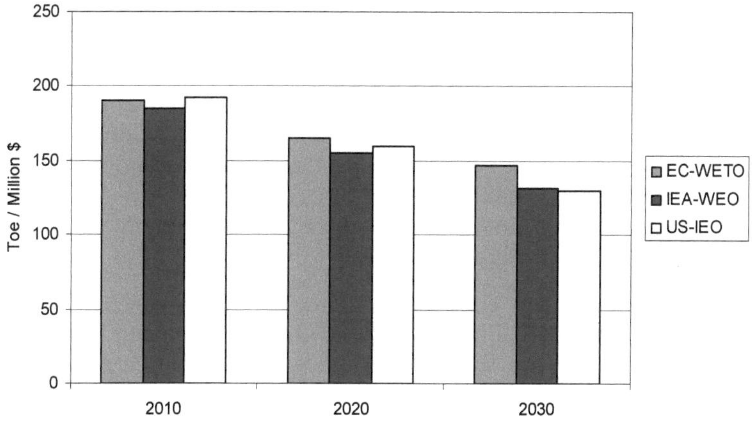

The world projected energy CO_2 emissions are quite similar among the three studies with about 28-29 Gt CO_2 in 2010, 33-34 Gt CO_2 in 2020 and 39 Gt CO_2 in 2030 in the EC-WETO and the IEA-WEO studies. The US-IEO results are about 15% higher mainly due to the strongest GDP growth assumptions (and consequent energy consumption) and due to the most important contribution of fossil fuels, especially coal in the American projections. Overall, the three studies expects a doubling of CO_2 emissions compared to 1990 (20.8 Gt CO_2), the reference year of the Kyoto Protocol.

Figure VII.11: Projected world CO_2 emissions in the EC-WETO, IEA-WEO and US-IEO studies (in Gt CO2)

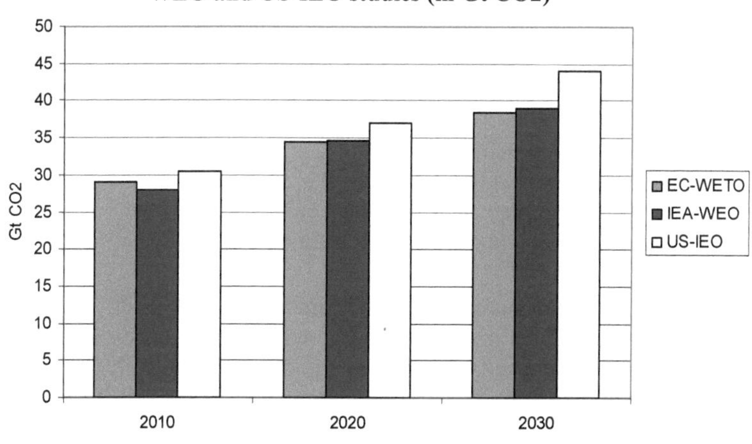

While total energy consumption and CO_2 emissions are very closely converging in the three compared scenarios at the 2030 horizon time (only a few percentage differences), larger discrepancies are noticed in the fuel mix of the total energy consumption.

In 2030, coal accounts for less than 24% in the EC-WETO study and more than 27% in the US-IEO study. The share of oil – that remains the first energy source by 2030 with more than 5 Gtoe – is relatively similar among the three studies. Natural gas has the most important share in the US-IEO study (27%) and the lowest share (22%) in the IEA-WEO study while it accounts for almost 25% in the EC-WETO report. It accounts for about 4 Gtoe in the three studies.

Large discrepancies concern the share of nuclear and renewables. Nuclear accounts for 8% in the EC-WETO study and about half of it in the US-IEO and in the IEA-WEO studies. Renewable energy sources and technologies (mostly biomass) account for less than 9% in the US-IEO study and more than 14% in the IEA-WEO study while they account for about 12% in the EC-WETO study.

Figure VII.12: Energy consumption fuel mix by 2030 in the EC-WETO, IEA-WEO and US-IEO studies (in percentage)

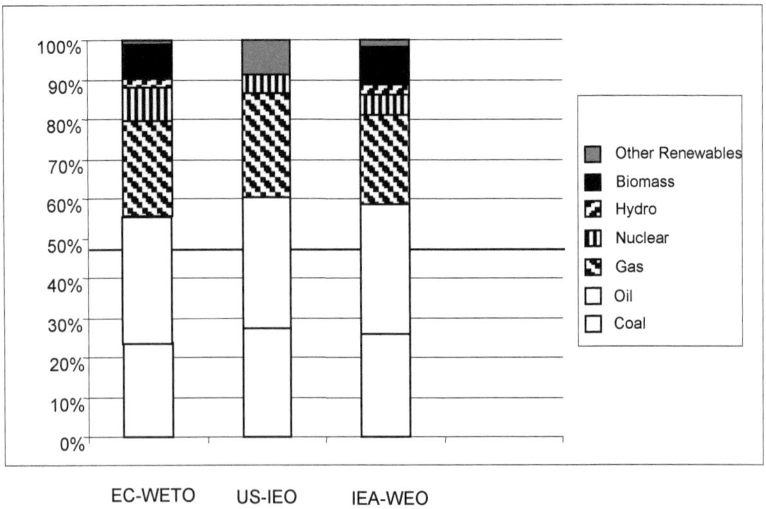

The differences on nuclear and renewables are even larger when looking at electricity generation. Whereas nuclear accounts for about 20% of world electricity generation in 2030 in the WETO-H2 study (more than 6000 TWh), it represents about 10% in the US-IEO and IEA-WEO studies. Concerning renewable electricity, its total share is very similar in the three studies (about 20%) but the classification is different. The US-IEO summarise all renewables (including biomass and hydropower) under "other renewables" and in the EC-WETO study, "hydro" includes geothermal. The problem of different definitions of biomass (commercial bioenergy and traditional use of energy in developing countries) also renders the distinction difficult.

Figure VII.13: Electricity generation fuel mix by 2030 in the EC-WETO, IEA-WEO and US-IEO studies (in percentage)

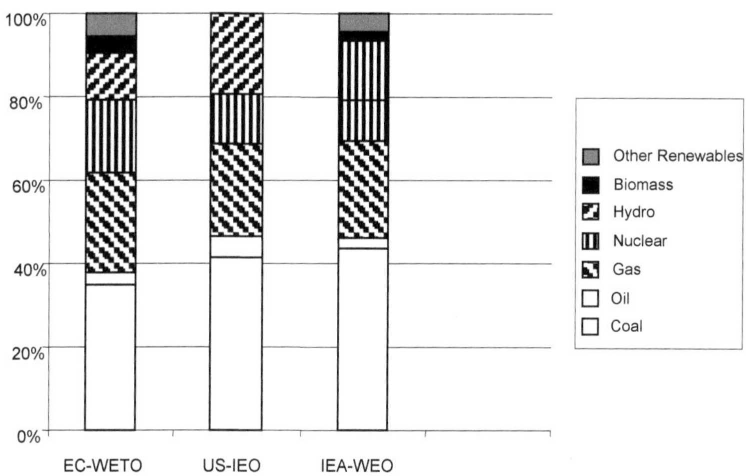

As a conclusion of this Chapter, two points deserve to be highlighted:

- Firstly, as there are only few comparisons between the realized and the forecast effects, the analysis of the ENERGY 200 study interestingly showed an impressive accuracy of prospective estimates in terms of the overall EC energy consumption. Nevertheless, as the models did not get the unexpected political, economic and social changes, for some fuels and technologies, large differences exist. For example, the share of natural gas in the EC electricity consumption has been largely underestimated by the forecast.
- Secondly, there is a clear concomitancy among the results of the three world energy outlook led by different teams using different Bottom-up models. Fossil fuels continue to be largely dominant by 2030. It is a typical "continuity foresight" (see Chapter III) without any technological breakthroughs or major socio-economic pattern changes.

CHAPTER VIII

EU Interaction of Science and Policy

Summary

The European Union has a *sui generis* decision-making process allowing in the same time to respect the general European interest and the national representation. Within the "Community method", the elected European Parliament and the Council (Member Sates) legitimately assume their role of co-legislator.

In this system and with this method, the European Commission has, among other roles, the monopoly of the right of initiative. As such, its legitimacy seems mostly grounded on its technical and scientific capacities as demonstrated in Chapter VI by its intense use of quantitative tools in the draft EU legislation also called "soft law".

The various Commission Communications, Green or White Papers – published in the Official Journal but without having a legal value – are an essential part of the "new legislative culture" highlighted in the European Governance in 2001. In this up-stream decision-making process, including the Commission Staff Working documents, the role of scientists (economists and modellers in the case of energy-related policies) is apparent and extremely strong.

While science (and technology) starts to be challenged by society, the role of scientists in EU policy-making is being more and more formalised. The "scientific support to policies" extends from the primary European legislation, i.e. the EU Treaty (already since 1986 for environment policy in the Single European Act) to the principles and guidelines on the use of expertise in the EU passing by the good governance principles.

In a more concrete way, the role of scientists in shaping EU policies can be found in the increased role of experts in the various committees that accompany the Commission in the execution of delegated powers ("Comitology") and in the formal procedure of Impact Assessment that is required for any important proposal presented by the Commission like regulatory proposals or expenditure programme. Impact assessments are now a key part of the policy making culture of the Commission.

From the previous chapters, three necessary parameters are deduced to explain the use of scientific tools in EU policy-making process. The models should be scientifically peer-reviewed; they should be well-managed and provide relevant policy work; and they should be subject to a long-time acquaintance by the Commission, Member States experts and stakeholders (industry, trade-unions and NGO).

In terms of timing, from the theoretical and empirical analyses, one can conclude that from the initial development of the tool (model) to the implementation of the new scientifically supported policy (greenhouse gas, renewable or energy efficiency targets, level of energy taxation or internalisation of external costs), ten years are needed trough a long process of academic, political and stakeholder validation where the Commission plays a key role as a "techno-political institution".

European decision-making and the role of scientists

Major principles of EU decision-making process

Several authors have described the EU decision-making process in general or focusing on the functioning and mechanisms of its main political institutions, i.e. the European Council, the European Parliament or the European Commission[1].

The objective of this section is not to provide a new analysis of the EU political system but, in a first step, to highlight its specificities and, in a second step, to detail why, how and when scientists (experts, technicians) are involved in EU decision-making process.

Regarding the specificities of the EU political system, at least the following seven points deserve to be noted:

First of all, since the Schuman Declaration of 9 May 1950 until the constitutional Treaty signed on 29 October 2004 in Rome, the Heads of State and Government have had never unanimously referred explicitly neither to a "federation" neither to a "confederation". Member States – six, nine, ten, twelve, fifteen, twenty-five or twenty-seven today – have always used a more vague terminology like European "Community" or European "Union" to define this original political and institutional

1 L. CARTOU *et al.*, *L'Union européenne*, *op. cit.*; H. WALLACE *et al.*, *Policy-making in the European Union*, *op. cit.*; F. CHALTIEL, *Le processus de décision dans l'Union européenne*, *op. cit.*; N. MOUSSIS, *Accès à l'Union européenne*, *op. cit.*; J. RICHARDSON, *European Union, power and policy-making*, *op. cit.*; M. WESTLAKE, D. GALLOWAY, *The Council of the European Union*, John Harper, UK, 2006; R. CORBETT, F. JACOBS, M. SHACKLETON, *The European Parliament*, John Harper, UK, 2007; D. SPENCE, *The European Commission*, John Harper, UK, 2006.

system that the former Jacques Delors called "unidentified political subject" (*Objet Politique Non Identifié*).

Secondly, the various Treaties – of Paris (1951), of Rome (1957), of Maastricht (1992), of Amsterdam (1997), of Nice (2000) and of Lisbon (2007) – present major similarities with the Constitution of the Member States. One finds there the provisions concerning the territory, the organisation of the powers, the major freedoms and the methods of revision of the founding text. Result of the work of the European Convention, The Constitutional Treaty of 2004 goes even further e.g. by incorporating the Charter of Fundamental Rights or by proposing a long-term President of the Council and a European Minister for Foreign Affairs.

Thirdly, several Member States competences passed to the EU. It is the case, for example, for the establishment of the internal market (measures to remove the obstacles to free movement of goods, persons, services and capital), for the measures regarding the common commercial policy, the monetary policy for the countries of the Euro area, the general rules of competition, the common organisation of the agricultural markets or some essential elements of the transport policy. Apart from these "exclusive competences" – of attribution – of the EU, others competences are "shared" between the EU and the Member States. It is the case, for example, for environment, for research and development or for cohesion policy. Some other EU competences aim to "support" – to supplement – the Member States (culture, industry, tourism, education, etc).

Fourthly, the subsidiarity concept – initially mentioned in the "Spinelli Report" (project of European Constitution adopted by the European Parliament on 14 February 1984) – was included in the Single European Act and taken again in the Maastricht, Amsterdam, Nice and Rome II Treaties. Subsidiarity means that the EU assumes the tasks that Member States cannot exert separately or that they would exert with less effectiveness. Accordingly, the action of the EU is subsidiary to that of the Member States. Powers, competences, duties and obligations are distributed according to their greatest effectiveness and to the citizens' proximity. In addition to subsidiarity, the concept of proportionality means that all actions and measures taken by the EU have to be adapted to the needs and specific character of the various situations[2].

Fifthly, in accordance with the case law (*jurisprudence*) of the European Court of Justice, the Community law takes precedence over the

---

2 The principles of subsidiarity and proportionality (Article 5 of the Nice Treaty) have been put forward at the Edinburgh European Council (December 1992) as major foundations of the new EU legislative policy.

national law. In addition, the Luxembourg Court embodies not only the international jurisdiction functions but it also guarantees the uniform interpretation of the Community law (like the *Cour de Cassation* in some Member States), of the legality of the regulations (like the Council of State in certain countries), and of the respect of the Constitution by the legislators (Constitutional Court).

Sixthly, European integration exceeds today the "sectoral-level": coal and steel, atomic energy, customs barriers. Exclusively national prerogatives – like beating the currency – recently passed to the European level through the monetary union and the higher responsibility for the European Central Bank within the European System of the Central Banks (ESCB). Moreover, the EU has a certain financial autonomy thanks to its income and in particular thanks to its "own resources" (proportion of VAT, customs duties, agricultural levies).

Lastly, the EU comprises an internal and permanent dialectic between the centrifugal forces (the Member States) and the centripetal forces aiming to strengthen the power of the common institutions, in particular the European Parliament and the European Commission. Since the beginning of European integration, this dialectic has been present. Member States have always wanted to keep their prerogatives while the European institutions tried to take of the breadth. The case of the European Parliament is obvious: from an initial consultative power, it obtained a budgetary power, a cooperation power and gradually a real "co-decision" power.

Role of scientists in EU-decision-making process

One of the objective of this book is to demonstrate the role of scientists (academics and experts or, in the case of energy-related policies, economists and in particular modellers). This role has been largely underlined by the five case-studies in Chapter VI. But some general aspects of the EU decision-making system have to be highlighted in order to understand why, how and when scientists can be so deeply involved in EU decision-making process.

The first point is related to the "Community method" imagined by Jean Monnet and Robert Schuman since the beginning of the European integration. This method aims to guarantee at the same time, the diversity and the effectiveness of the EU. It ensures the equitable treatment of all the Member States, from the largest to the smallest. It provides a means of arbitrating between various interests through two successive filters: the filter of the general interest at the level of the Commission; and the filter of the democratic, European and national representation at the level of Parliament and of the Council, which constitute together the legislative power of the EU.

With the Community method, the European Commission is the only one to formulate legislative and policy proposals (monopoly of the right of initiative). Its independence enhances its capacity to make policy, to act as guardian of the Treaty and to represent the Community in international negotiations. The legislative and budgetary acts are adopted by the Council of Ministers (representing the Member States) and the European Parliament (representing the citizens). The use of the vote by qualified majority in the Council is an essential element to ensure the effectiveness of this method. The implementation of policies is entrusted to the Commission and to the national authorities. The European Court of Justice guarantees the uniform interpretation of Community law.

Although the European Commission is one of the three EU political institutions, it neither represents the "national mandates" (Members States) as the Council nor the European citizens like the elected European Parliament. As demonstrated in Chapter VI with the energy-related policy case studies, one can say that the "legitimacy" of the European Commission has to be found in its "technical" character[3]. The legislative right of initiative of the Commission is most of the time grounded on scientific and technical work. It is not an accident that so many observers speak about the European technocracy[4].

Without being democratic like the Parliament (elected since 1979) and without assuming the legitimacy of the Council (national Ministers), the Commission uses science, facts and figures, to underpin its authority, in particular when making policy proposals. If exact sciences results are not easy to refute, this is not the case in social sciences like economics. Therefore, as shown in the analysed energy-related policy case-studies (Chapter VI), the Commission has the double tendency:

- To use more quantitative methods than qualitative ones, probably because the quantitative results seem more "exact" and less refutable. According to the Commission, classical scientific assessments

3 Several academic papers focus on a specific EU legitimacy for example "social legitimacy" (J. WEILER, *Problems of legitimacy in post 1992 Europe*, Aussenwirtschaft 46, 1991) or "legitimacy granted by the rule of law" (M. FERNANDEZ ESTEBAN, *The rule of law in the European Constitution*, Kluwer, The Hague, 1999) or "substantive legitimacy" as underlined in many occasions by Pierre Pescatore that served the European Court of Justice during the 1970s and the 1980s. More recent papers attempt to demonstrate that the EU stands on a plurality of legitimating principles: F. SCHARPF, *Notes toward a theory of multilevel governing in Europe*, Max-Planck-Institut für Gesellschaftforschung, Discussion Paper, 5/2000; C. LORD, P. MAGNETTE, *Notes towards a general theory of legitimacy in the European Union*, ESRC Working Paper 39/02, Sussex European Institute, 2001.

4 C. RADAELLI, *Technocracy in the European Union*, Longman, London, 1999.

– in the natural or social sciences – are often "uncontroversial" (see next section of this Chapter)[5].

- To look for "scientific consensus results" although the concept of "consensus" itself is more political than scientific. Often, in making the synthesis of the expert groups, the Commission employs the expressions like: "members of the Group agreed" or "consensus was achieved among Member States experts"[6].

The second point, also underlying the "hidden role of scientists" in EU decision-making can be observed in the typology of European legal acts. The European Treaties are at the top of the legal pyramid. They are the sources of the Community law. The secondary legislation includes the Directives, the Regulations and the Decisions, i.e. the legal acts adopted based on an article of the Treaty[7]. The execution measures – that can take the form of Directives, Regulations or Decisions – are the application rules of the secondary legislation. The "soft law" is at the bottom of the EU legal act typology. It includes the legal acts that are not binding, i.e. the Recommendations, the Inter-institutional Agreements, the Rules of procedure and the Guidelines. One could also add the "unidentified European legal acts" – i.e. the Commission Communications, Green and White Papers – that are not clearly and unanimously categorised under "soft law"[8].

Since the middle of the 1980s, the EU legislative leitmotiv was going from "binding harmonisation legislation" to "non-binding coordination of national policies". Ten years later, the European Commission was speaking about a "new legislative culture"[9]. In 2001, the Commission raised the importance of soft law in at least two opportunities: in the White Paper on European Governance (see next section of this Chapter) and in the Communication on taxation.

---

5 European Commission, *Communication on the collection and use of expertise by the Commission: Principles and guidelines – Improving the knowledge base for better policies*, COM(2002)713.

6 European Commission, DG TAXUD, *Common Consolidated Corporate Tax Base Working Group*, Progress to date and future plans for the CCCTB, Brussels, 20/11/2006.

7 Regulations are binding in their entirety and are directly applicable in all Member States. Directives bind the Member States as to the results to be achieved and leave a margin for manoeuvre as to the form and means of implementation; Concretely, a EU Directive is a type of law that binds Member States to the result but allows flexibility as to how it is implemented. Decisions are fully binding on those Member States to whom they are addressed.

8 L. SENDEN, *Soft law in European Community law*, Hart Publishing, UK, 1994.

9 Report on implementation of the Commission's work programme for 1996, European Commission, Brussels, 16 October 1996.

The European Governance Communication stated "that the EU must renew the Community method by following a less Top-down approach and complementing its policy tools more effectively with non-legislative instrument". It added that "legislation is often only part of a broader solution combining formal rules with other non-binding tools such as recommendations, guidelines, or even self-regulation within a commonly agreed framework"[10]. In the Communication on taxation, the Commission explicitly recognised that the use of non-legislative approaches or soft legislation may be an additional means of making progress in the tax field[11].

As demonstrated in Chapter VI on the energy-related policy case studies, the impact of the scientists (modellers) are mainly used in the European "soft law", i.e. in the Communications, Green and White Papers. In addition, the scientists results are most of the time reflected in the Commission Staff working papers and in the Impact Assessments (see also the next section of this Chapter), legal acts that have not yet been categorised by lawyers[12].

At the preparatory stage of a European decision, the research work is clearly recognised and directly used. In the next step of the EU decision-making process (from soft-law to execution measures or secondary legislation), the scientists results are not anymore quoted. *De facto*, not one scientific result is mentioned in the Directives analysed in the five energy-related policy case-studies. When passing from Commission Staff document or from "Soft law" to Directive, the research works are in a certain sense technically absorbed and politically validated by the Commission, the "most techno-political European institution".

---

10 European Commission, Communication, *European Governance – a White paper*, COM(2001)428.

11 European Commission, *Tax policy in the European Union – Priorities for the years ahead*, COM(2001)260.

12 In the European Official Journal (OJ) are published the L series (Legislation), the C series (Communications – information, preparatory acts and notices) and the S series (Supplement – calls for tenders). The Commission Staff working papers (SEC documents in the European jargon) are not published in the OJ.

Figure VIII.1: Decision-making process in the European Union including the role of scientists

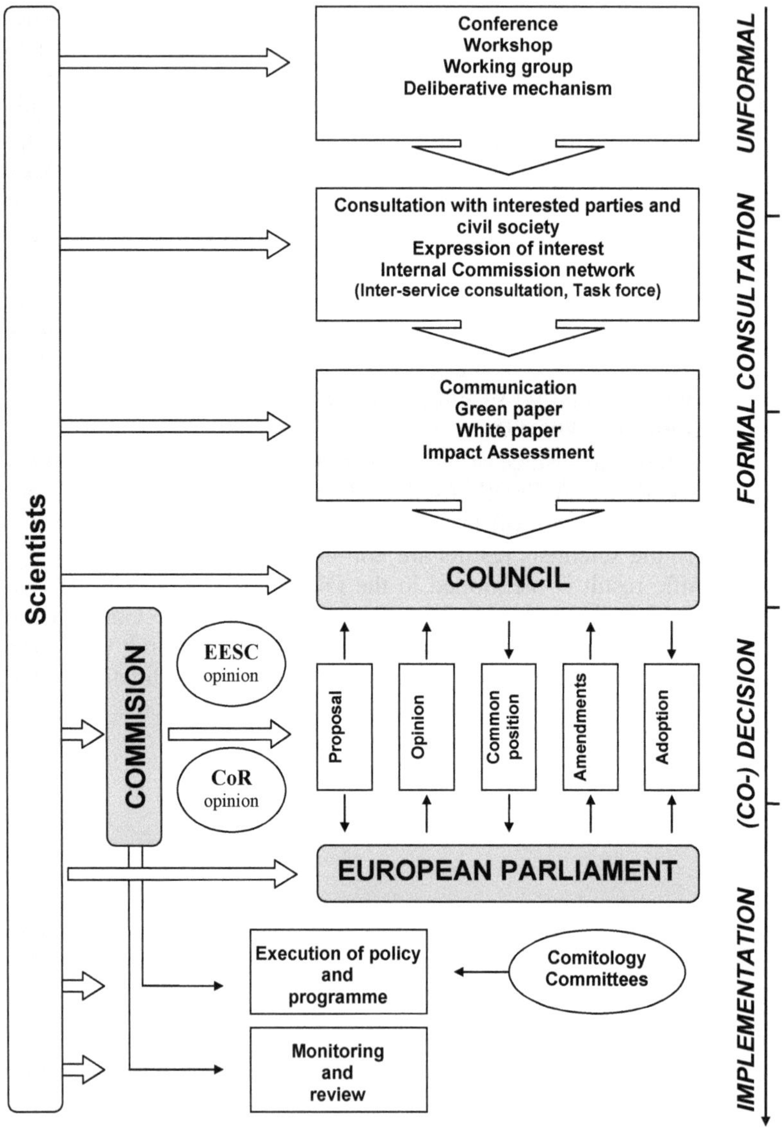

Legal formalisation of the influence of scientists in the EU

From Jean-Jacques Rousseau to Chris Freeman and Luc Soete while passing through Jacques Ellul and Martin Heidegger, from the 18th century to most recent days, Science and Technology (S&T) have been largely discussed from several discipline points of view but only rarely connected with policy-making[13].

At the end of the 1990s, Chris Freeman and Luc Soete highlighted the key role of S&T and innovation in economic development. They tackled both the micro-economic and the macro-economic aspects of innovation through the theory of the firm and the impact of S&T on economic growth. They demonstrate the central role of knowledge and of information and communication technologies as competitive strengths (greater productivity) in the context of globalisation (see Chapter IV)[14]. Although recognising the important role of technological development, Jean-Hervé Lorenzi and Jean Bourlès are more balanced, in particular in relation to the employment issue:

> Il y a la conviction, excessive à nos yeux, que la technologie est la variable déterminante (…). L'innovation technologique est apparue successivement dans les vingt dernières années comme porteuse d'espérances excessives, puis à la source de tous les maux[15].

---

13 In 1750, J.J. Rousseau took historical examples to demonstrate that sciences and arts were responsible for the weakening of the man and for the decline of the manners. Knowledge was often considered as vanity and, for Rousseau, education – other than learning virtue – was risky. J.J. ROUSSEAU, *Discours sur les sciences et les arts*, Paris, 1750.

According to J. Ellul, the key of the modernity is not anymore in the economics (like at the time of Marx) but in the technical factor. In the twentieth century, technique is not only an intermediary between man and its natural environment but an autonomous process responding to its own rules. J. ELLUL, *La technique ou l'enjeu du siècle*, Armand Collin, Paris, 1954; J. ELLUL, *Le système technicien*, Calman-Lévy, Paris, 1977.

M. Heidegger points out the dangers of S&T, from the most obvious to the most profound. S&T sum their objects which become simple objects of study to knowledge (for science) and simple raw materials (for technique). The relation is between a dominator subject and unauthentified objects. S&T submit everything to a dominating reason which searches manufactured effects. Both of them conquer and enslave the nature of one intellectually, the other materially. In the essential relationship from earth to things, S&T make man a stranger to himself. M. HEIDEGGER, *Überlieferte Sprache und technische Sprache*, Conference 18/7/1962, Comburg, Schwäbisch Hall, Erker Verlag, Sankt-Gallen, 1989.

14 C. FREEMAN, L. SOETE, *The economics of industrial development*, MIT Press, 1997.

15 J.-H. LORENZI, J. BOURLES, *Le choc du progrès technique*, Economica, Paris, 1995.

Since these last forty years, especially with the questions of health, food safety, environment and energy, there is a major change in social acceptability of S&T. Some energy issues raised high public concerns like the Torrey Canyon super tanker disaster in 1967, the nuclear accident of Chernobyl in 1986, the climate change issues publicly highlighted at the 1992 Rio Summit – and recently accentuated by the weather extremes in several parts of the world, the Erika oil spill in 1999 or the increased oil price also caused from geopolitical conflicts and tensions (Iraq, Iran) these last few years.

More and more authors note the necessity for society to control S&T. Jean-Jacques Salomon wrote:

> La technologie n'est plus l'habit contemporain de l'enchaînement nécessaire des choses (...). Limiter, réguler, contrôler le changement technique, réfréner son rythme, corriger sa direction, maîtriser sa nature: tel est l'enjeu du XXIe siècle[16].

Wiebe Bijker, Tom Hughes and Trevor Pinch raised the issue of opacity and lack of transparency in S&T. "Closure occurs in science when a consensus emerges that the 'truth' has been winnowed from the various interpretations"[17].

Michael Schrage underlined that

> the tragedy is not that science is too political – it is that science is not political enough (...). Public policy would be significantly better off if scientists were treated with greater scepticism and less deference (...). Politicians should not hesitate to exploit publicly the inherent uncertainties and legitimate disagreements in scientific analyses on sensitive issues (...). United Nations agencies have increasingly abandoned traditional roles as science 'advisers' in favour of actively lobbying for their quantitative models and scenario extrapolations to be public policy planning tools (...). The scientific 'objectivity' of their forecasts is achieved through negotiated committee consensus. Unfortunately, most of these consensus declarations minimise methodological disagreements, competing interpretations and self-criticism[18].

James Wildson, Brian Wynne and Jack Stilgoe emphasise the importance of a dialogue between experts and non experts in order to ensure that these latter acquire the kind of local knowledge, lay knowledge and lay expertise which contributes to a socially robust scientific view[19].

---

16 J.-J. SALOMON, *Le destin technologique*, Gallimard, Paris, 1992.

17 W. BIJKER, T. HUGHES, T. PINCH, *The social construction of technology systems*, MIT Press, Cambridge, 1987.

18 M. SCHRAGE, *Science must be more political*, Financial Times, 26/9/2007.

19 J. WILSDON, B. WYNNE, J. STILGOE, *We need to infuse the culture and practice of science with a new set of social possibilities – The public value of science or how to ensure that science really matters*, DEMOS, UK, 2005.

At the same time, when S&T start to be strongly challenged by society, the role of scientists (experts or "modellers" in this book) in EU decision-making process increase. From the end of the 1990s, at the EU-level, this has been "formalised" in four different ways – through the EU Treaty; the good governance principles; the scientific support to policies; and the principles and guidelines on the use of expertise in the EU. The role of scientists in EU decision-making has also been practically implemented mainly through the so-called "Comitology" and "Impact Assessment" (see next section of this chapter).

European Treaty

The Treaty establishing the European Community recognises the importance of facts and data in relation to health and safety, environment and consumer protection[20].

The article 95 (3) states:

> The Commission, in its proposals (...) concerning health, safety, environmental protection and consumer protection, will take as a base a high level of protection, taking account in particular of any new development based on scientific facts.

The article 174 (3) indicates that

> In preparing its policy on the environment, the Community shall take account of available scientific and technical data; environmental conditions in the various regions of the Community; the potential benefits and costs of action or lack of action; the economic and social development of the Community as a whole and the balanced development of its regions.

Beyond these articles of the Treaty, expert assessments underpin a large range of issues like authorising cosmetic products, establishing automobile safety standards, determining sustainable fish catches, developing strategies to tackle unemployment, or, as abundantly demonstrated in Chapter VI, for fixing energy-related policy targets like greenhouse gas emission reduction, share of renewable energy sources or energy efficiency improvements.

European Union good governance principles

Two out of the five principles of good governance indicated in the European Governance White Paper (openness, participation, accountability, effectiveness and coherence) refer, in a certain way, to scientific – and modelling – activities[21]:

---

[20] European Union, *Consolidated versions of the Treaty on EU and of the Treaty establishing the European Community*, OJ, C 321, 29/12/2006.

[21] European Commission, *Communication, European Governance – a White paper*, *op. cit.*

- Effectiveness, meaning that policies must be effective and timely, delivering what is needed on the basis of clear objectives, *an evaluation of future impact* and, where available, of past experience (cf. next section of this Chapter on Impact Assessment);
- Coherence of policies and action. The search for coherence is one of the main objectives pursued by modelling (systematic) analyses. This coherence is particularly relevant for environment and sustainable development (cf. the "integration" of environment into all EU policies since 1998 – "Cardiff process")[22].

Scientific support to EU policies

In 2000, one of the European Research Area (ERA) objectives was to develop a "Common system of scientific and technical reference for policy implementation". This included two main items:

- Development of the research needed for political decision-making: The results of research undertaken as part of European programmes should be systematically exploited in support of the various EU policies and all the EU research activities better co-ordinated in this respect[23].
- Establishment of a common system of scientific and technical references: A reliable and recognised system of validating knowledge and methods of analysis, control and certification should be put in place and centres of excellence in Europe in the field concerned networked[24].

In the sixth EU RTD Framework Programme (2002-2006), a programme called "Scientific Support to Policies"[25] was launched. It had the objective to "underpin the formulation and implementation of Com-

---

22 European Commission, *A strategy for integrating the environment into EU policies*, COM(1998)333; European Commission, *Integrating environmental considerations into other policy areas – a stocktaking of the Cardiff process*, COM(2004)394; OECD, *Improving governance for sustainable development*, Seminar, Paris, 22-23/11/2001.

23 European Commission, *Towards a European Research Area*, *op. cit.*

24 European Commission, *Making a reality of the European Research Area: guidelines for EU research activities*, *op. cit.* It is also worthwhile reminding that since 1957 the European Joint Research Centre (JRC) functions as a reference centre of S&T for the EU (See Chapter II), at the beginning mostly in the nuclear field. Today the nine institutes of the JRC provide scientific information useful for the design, implementation and assessment of Commission services' policies. The JRC also aims to produce socially robust knowledge and thereby contribute to enhancing the credibility and legitimacy of science inputs in public policy and social discourse.

25 European Parliament and Council, *Decision concerning the FP6 of the EC for RTD&D, contributing to the creation of the ERA and to innovation (2002-2006)*, OJEC, 29/8/2002.

munity policies, bearing on the interests of possible future members of the EU as well as the existing member states, and monitor their effects"[26]. Directly related to energy, the programme dealt with the development of tools, indicators and operational parameters for assessing sustainable transport and energy systems performance – economic, environmental and social[27]. In order to promote a more efficient use of scientific information and expertise in support to policy making, the European Commission developed a web communication platform called SINAPSE.

In 2007, the ERA Green Paper highlighted that European research policy should be deeply rooted in European society[28]. Besides the pursuit of scientific excellence, European research should support knowledge advancement and dissemination and underpin policies for sustainable development in fields of major public concern such as energy and climate change. It stated that innovative approaches are needed to improve the availability and uptake of relevant S&T expertise for evidence-based policy-making.

Principles and guidelines on the use of expertise in the EU

The idea that "the Commission needs expertise to ensure that EU policies have a sound knowledge base" is explicitly mentioned in the Principles and Guidelines on the collection and use of expertise by the Commission services in 2002[29]. As the expertise needed is wide and the knowledge required is increasingly technical and highly specialised, the Commission calls upon external specialists in different fields to provide advice.

Expertise may take several forms, often including scientific knowledge and sometimes practical knowledge. Experts are asked to perform "classical" scientific assessments, rooted in the natural or social sciences. According to these Principles and Guidelines the expertise is often a straightforward process in which the questions to be addressed, the evidence considered and the interpretations made are "uncontrover-

---

26 Council *Decision adopting a specific programme for research, technological development and demonstration: Integrating and strengthening the ERA (2002-2006)*, OJ, L 294/1, 29/10/2002.

27 European Commission, *Work Programme for the specific programme for RTD&D "Integrating and strengthening the ERA" – Specific activities covering a wider field of research, Policy support and anticipating scientific and technological needs*, 9/12/2003.

28 European Commission, *The European Research Area: New perspectives*, *op. cit.*

29 European Commission, *Communication on the collection and use of expertise by the Commission: Principles and guidelines – Improving the knowledge base for better policies*, *op. cit.*

sial". On other occasions, however, the Commission recognises that some aspects – like GMOs or the avian influenza – are highly controversial and that difficult policy decisions must sometimes be made on contentious issues in the face of large uncertainty.

Three principles underpin the Commission activities in the use of expertise:

- Quality: The determinants of which are excellence; the extent to which experts act in an independent manner; and pluralism;
- Openness: This imply transparent procedure (particularly in relation to the way issues are framed, experts are selected, and results handled) and proactive communication;
- Effectiveness: Arrangements and resources for collecting and using expertise should be designed in proportion to the task in hand, taking account of the sector concerned, the issue in question, and the stage in the policy cycle.

The guidelines concern the necessity to plan; the preparation for the collection of expertise; the identification and selection of experts; the management of the involvement of experts; and the way to ensure openness.

The Figure below – coming from the Principles and Guidelines on the collection and use of expertise – highlights the imbrication among the Commission, Experts and Interested parties (stakeholders) that reflect the "new legislative culture" above-mentioned.

Figure VIII.2: Commission consultative process of experts and interested parties

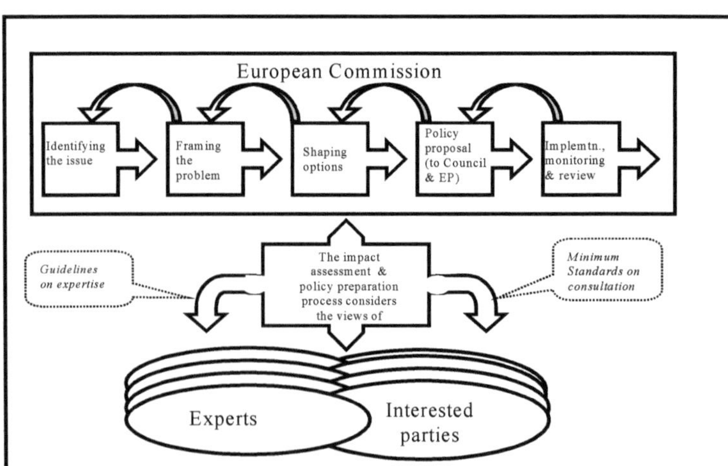

Applied role of experts in shaping EU policies

Two majors aspects reflect the increased role of scientists (experts) in day-to day EU policy-making: firstly, the increased weight of "Comitology" and "Expert Working Groups"; and secondly, the formal necessity of providing an "Impact Assessment" for all major proposed EU policies.

Comitology and expert groups

The Article 202 of the European Treaty states that, "the Council shall confer on the Commission, in the acts which the Council adopts, powers for the implementation of the rules which the Council lays down"[30].

Comitology refers to the system of committees that control the Commission in the execution of delegated powers. In these Commission expert committees, Council working groups or "Commission implementation comitology committees", national representatives and experts participate in shaping the European decisions. The formal committees – initially developed in the 1970s for the implementation of the Common Agricultural Policy (CAP) – are today more than 250. In addition, hundreds of *ad hoc* working groups are part of the EU implementing powers[31].

The issue of "democracy" or parliamentary scrutiny over Comitology has been largely accentuated by the increased co-decision power of the European Parliament (EP) since the Single European Act (1986), the Maastricht, the Amsterdam and the Nice Treaties (1991, 1997 and 2000)[32]. The EP was increasingly dissatisfied with the exclusive arrangement for Member State representatives to control the Commission. The EP was requesting to be more involved in this process. Although the Council Comitology Decisions of 1987 and 1999 – and the inter-institutional agreements – addressed this issue, in 2006 the Council

---

30 European Union, *Consolidated versions of the Treaty on European Union and of the Treaty establishing the European Community, op. cit.*

31 C. JOERGES, J. NEYER, *From intergovernmental bargaining to deliberative political processes: the constitutionalisation of comitology*, European Law Journal, Vol. 3, No. 3, September 1997; J. WEILER, *Epilogue: Comitology as revolution – Infranationalism, constitutionalism and democracy*, in C. JOERGES, E. VOS (eds.), *EU Committees: Social regulation, law and politics*, Hart Publishing, Oxford, 1999; B. STEUNENBERG, B. KOBOLDT, D. SCHMIDTCHEN, *Comitology and the balance of power in the European Union*, International review of Law and Economics, 1996; W. WESSELS, *Comitology: fusion in action – Politico-administrative trends in the European Union system*, Journal of European Public Policy, Vol. 5, No. 2, 1998; T. CHRISTIANSEN, T. LARSSON, *The role of committees in the European-Union policy-process*, Edward Elgar, 2007.

32 M. RHINARD, *The democratic legitimacy of the European Union committee system*, Governance, Vol. 15, April 2002.

decision adopted the 17th of July provided a more acceptable solution for the EP. A new "regulatory procedure with EP scrutiny" was created for "quasi-legislative act" that arise from co-decided basic acts[33].

De facto, the rights of the EP were recognised for these "measures of general scope designed to amend non-essential elements *inter alia* by deleting some of those elements or by supplementing the instrument by the addition of new non-essential elements" as stated in the 2006 Decision ("quasi-legislative" implementing measures)[34]. The EP has the right to veto the adoption of these measures[35].

The influence of experts in "technical working groups" for drafting EU policies or for evaluating – and therefore funding – European activities is also substantial. Since its origin in the 1950s, the whole European Economic Community was largely considered as a "technocracy" because it dealt mainly with a technical issue, the establishment of the common market.

The following six examples help to illustrate the influence of these different expert working groups:

Since the 1960s up to now, many national and expert committees are managing large parts of the implementation of the Common Agricultural Policy (CAP). Agricultural legislation accounts for over half of all Community legislation and the management/regulatory/advisory committees are consulted on some 2000 texts per year[36].

In the trade policy (an exclusive competence of the Commission and where the Commission negotiates on behalf of the Member States), the "Committee 133" – that meets weekly in Brussels – deals among others with the European position at the World Trade Organisation (WTO). Before the entry into force of the Amsterdam Treaty, this Committee was called "Committee 113" from the original article of the Rome Treaty[37].

The European Central Bank (ECB) draft statute adopted by the European Council in Masstricht (1991) reflected – almost word for word

---

33 European Parliament, Council, Commission, *Inter-institutional Agreement on better law-making*, OJ 2003, C 321/01. This Agreement completes the 1994, 1998 and 2001 Inter-institutional Agreements.

34 Council *Decisions 87/373/EEC, 1999/468/EC, 2006/512 laying down procedures for the exercise of implementing powers conferred on the Commission*, OJ L 197, 18/7/1987, OJ L 184, 17/7/1999, OJ L 200, 22/7/2006.

35 CHRISTIANSEN T., VACCARI B., *The 2006 reform of comitology: Problem solved or dispute postponed*, EIPASCOPE No. 3, Maastricht, 2006.

36 European Commission, *Agricultural Committees*, Brussels, 2000.

37 S. MEUNIER, N. KALYPSO, *Who speaks for Europe? The delegation of trade authority in the EU*, Journal of Common Market Studies, Vol. 37, No. 3, September 1999.

– the text of a "technical committee" composed of central bankers. The discussed independence of the ECB is also considered by many observers as associated to a "technical matter", the European Monetary Union[38].

Several experts groups are also created for specific policies, often where unanimity is required like taxation. This is the case for example for the Common Consolidated Corporate Tax Base Working Group (CCCTB) that includes experts from the 27 Member States[39].

In the EU Research policy (€ 55 Billion for the seventh Framework Programme – 2007-2003), the projects submitted following a Call for Proposals are evaluated by peer-review, i.e. by experts. In almost all the cases, the Commission follow precisely the recommendation of the experts in its final selection decision (see Chapter II).

Figure VIII.3: Role of experts in the selection of EU research proposals

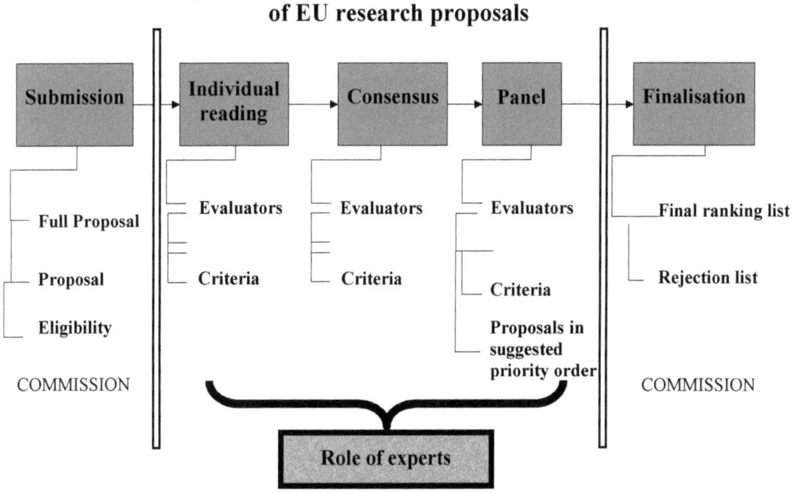

Impact assessment

The formal Impact Assessment founds its origin in two basic concepts, one related to "Better regulation" and one to "Environment and Sustainable development".

38 K. DYSON, K. FEATHERSTONE, *The road to Maastricht: Negotiating Economic and Monetary Union*, Oxford University Press, 1999.

39 European Commission, DG TAXUD, *Common Consolidated Corporate Tax Base Working Group, op. cit.*

The Laeken European Council (2000) and the Seville Council (2002) asked to introduce Regulatory Impact Assessment as part the Better Regulation Action Plan[40]. The idea was to assess the alternative options for EU proposals (Regulation, Directive, Voluntary agreement etc.), to examine the case for Community action (subsidiarity) and to examine the extent to which the EU should act (proportionality). "Better Regulation covers policy making, from its initial conception to implementation and enforcement starting with the careful application of the principle of subsidiarity"[41].

The 2001 Göteborg European Council asked the Commission to introduce a Sustainable Development Impact Assessment that would assess the impact of proposals on the economy, on the natural environment and on the society[42]:

> The EU Sustainable Development Strategy sets out an approach to better policy-making based on better regulation and on the principle that sustainable development is to be integrated into policy-making at all levels. This requires all levels of government to support, and to cooperate with each other, taking into account the different institutional settings, cultures and specific circumstances in Member States. In this respect all EU institutions should ensure that major policy decisions are based on proposals that have undergone high quality Impact Assessment, assessing in a balanced way the social, environmental and economic dimensions of sustainable development and taking into account the external dimension of sustainable development and the costs of inaction. Other tools for better policy-making include *ex-post* assessment of policy impacts and public and stakeholders participation. Member States should make wider use of these tools, in particular Impact Assessment, when allocating public funds and developing strategies, programmes and projects[43].

The objective of the Strategic Environmental Assessment Directive (cf. article 1) is:

> To provide for a high level of protection of the environment and to contribute to the integration of environmental considerations into the preparation and adoption of plans and programmes with a view to promoting sustainable

40 European Commission, *Simplifying and improving the regulatory environment*, COM(2002)278. The origin of "Better regulation" find its origin in the conclusions of the Edinburg Summit of December 1992. "Better" is also often accompanied by "Less regulation" considering that the current *Acquis Communautaire* represents today about 100,000 pages in the *Official Journal*.

41 European Commission, *A strategic Review of better regulation in the European Union*, COM(2006)689.

42 European Commission, *A sustainable Europe for a better world: A European strategy for sustainable development*, COM(2001)264.

43 Council of the European Union, *Review of the EU sustainable development strategy – Renewed strategy*, 9/6/2006.

development, by ensuring that, in accordance with this Directive, an environmental assessment is carried out of certain plans and programmes which are likely to have significant effects on the environment[44].

In 2002, the Commission presented its Communication on Impact Assessment (IA)[45]. From 2003, an IA is required for any important proposal presented by the European Commission. An IA aims to identify the likely positive and negative impacts of proposed policy actions, enabling informed political judgements. Its main objective is in line with the European Governance White Paper[46]. In particular, it should help to provide "Good Governance" in terms of sustainability and transparency of policy-making.

IA is intended to improve policy design, i.e. making sure that the objectives are achieved in ways that maximise benefits whilst minimising costs. IA aims to improve quality and coherence of policy proposals (to ensure co-ordination early on in the policy design process and to combine existing diverse and separate IA into a single IA for all policy areas)[47]. As such, IA plays an integration role by ensuring that impacts from other policy areas are taken into consideration. Therefore, IA is potentially a key tool for making Sustainable Development a reality.

IA allows to assess the economic, social and environment impacts, to identify "gainers" and "losers", to provide qualitative and quantitative information (if possible also in monetary term) and to examine possible compliance problems and to consider risks and uncertainties. It helps policy-makers that are often in a complex situation with effects, side-effects and counter-effects with the necessity to manage in the same time global strategies or goals and sectoral measures or competences.

IA is based on four principles: Integration: global approach (with economic environmental, social impact, i.e. replacing all existing IA like Business Impact or Environmental IA); Co-ordination (IA promotes cross-sectoral approach); Proportionality (analysis match efforts to significance of impacts and assess if EU action is really needed); and Transparency (with the consultation of stakeholders and the publication of results).

---

44 European Parliament and Council *Directive 2001/42/CE on the assessment of the effects of certain plans and programmes on the environment*, OJ L 197, 21/07/2001.

45 European Commission, *Communication on Impact Assessment*, COM(2002)276.

46 European Commission, *Communication, European Governance – a White paper*, COM(2001)428.

47 In the previous chapters, several examples of IA of policies have been provided. But IA are also made for programmes or financial initiatives. A good example is the IA of the Hydrogen and Fuel Cell Joint Technology Initiative (€ 500 millions of EU funding), IA that has been largely based on the HYWAYS European research project that included a large part of modelling.

Table VIII.1: Scientists involvement in EU-decision-making process through Impact Assessment

Legal documents subject to IA	Legal documents not subject to IA
Regulatory Proposals	Green Papers
White Papers	Periodic Commission Decisions
Expenditure Programmes	Periodic Reports
Communications	Technical Updates
Negotiating Guidelines	Implementing Decisions

In practical terms, the IA is based on a two-stage process: a preliminary assessment (about two pages to identify proposals that need extended impact assessments) and an extended assessment (about twenty pages and making a thorough assessment of regulatory options and economic, environmental and social impacts).

The iterative process of IA can be divided into eight major questions:

- What problem is the policy/proposal expected to tackle?
- What are the risks inherent in the initial situation and what would happen under a "no-policy change" scenario (trade-offs associated with the proposed option)?
- What main objective is the policy/proposal expected to reach – considering that the objectives should describe the key required results within a target timeframe?
- What are the main policy options available (standards, taxation, voluntary agreements, emission trading) to reach the objective?
- What are the direct and indirect impacts – positive and negative, "gainers" and "losers" – expected from the different options identified?
- What are these impacts in qualitative and quantitative terms and monetize where possible (effects in comparison to a no-policy change scenario)?
- How to monitor and evaluate the results and impacts of the proposal after implementation – knowing that the quality of monitoring depends on the quality of the objectives?
- Which stakeholders are to be consulted taking into account that the "Commission should consult widely before proposing legislation"[48]?

With this IA, the Commission draft a proposal and its justification. Formally speaking, the IA is a "Commission Staff document". In its

48 European Commission, *Towards a reinforced culture of consultation and dialogue – General principles and minimum standards for consultation of interested parties by the Commission*, COM(2002)704.

elaboration, the IA requires the involvement of the Commission official responsible for the policy, the analyst involved in producing the assessment and the external stakeholders. From 2003 to 2006, the Commission has completed over 170 IA.

All impact assessments must include public consultations as part of the preparation process. Stakeholders play an essential role in providing necessary information, both qualitative and quantitative, to the Commission services.

In 2008, the Commission launched a consultation on the new Impact Assessment Guidelines and reinforced several elements: subsidiarity, proportionality, and the added value of a EU-initiative; risk assessment; social impacts; consumer impacts; impacts on small businesses; impacts at national and regional level impacts on international trade and investment, and on third countries; impacts on administrative burden and simplification.

Research work and policy implementation

From the previous chapters, it seems possible to deduce the three conditions that must be satisfied if research tools and methods are to offer effective support to policy-making.

Scientific peer-review

The models and tools – like POLES, PRIMES, GEM-E3, SAFIRE, MURE and EXTERNE – lengthily quoted in Chapter V and VI – were initially developed under funding from the Commission Research Directorate General (DG) and through the Call for Proposals procedure.

With such a procedure, the scientists have a large freedom for developing their tools. As they answer to a general Commission-written Work Programme, scientists can choose their methodology, their scientific focus, their assumptions and their empirical studies. In practical terms, they also propose their consortium and the research duration.

The proposal is evaluated by a peer-review system meaning that other researchers i.e. experts specialised in the field (generally at least three from different backgrounds, disciplines and citizenships) make the recommendation to the Commission on the best proposal to fund. The peer-review process is firstly done individually and later collectively (each expert make its own judgement and then meet with his/her colleagues to get a consensus). On average, in the energy economic research, one out of four submitted proposals is finally selected for funding.

At this stage, the methods and tools (models) can start to be developed but remain in the academic sphere (articles, publications, etc.) and have not yet policy applications. This stage is extremely important do develop the tool and to test the reliability of the model (completeness of the database, of the geographical and technological coverage, functioning of the models, first modelling runs).

Well-managed and policy-relevant work

When they have a specific policy needs, the political Directorates-General (DG Transport and Energy, DG Environment, DG Taxation, etc.) publish a Call for Tender. In this case, the description of the tasks is very detailed in a "Technical annex" accompanying the Tender.

The freedom of scientists is quite limited in terms of methodology to be used, of policy scenario to be addressed and of execution of the work. The required tasks are here more connected to "consulting" than to purely academic work.

The Call for Tender is directly evaluated by Commission civil servants (generally at least three from different Directorates-General). In practice, the "quality" and the "capacity of the consortium to achieve the expected results" are the two predominant criteria in both Calls for Proposals and Call for Tender. Nevertheless, from experience, one can say that the tendency of the peer-reviewers is to give more weight to the scientific quality while the Commission civil servants are inclined to consider as an essential aspect the ability of the consortium to answer in due time to the public tender specifications, in particular to the policy that the Commission want to initiate[49].

Durable Commission and stakeholder's acquaintance with the tools

As previously shown (see Figure VIII.1), the two first steps of the EU decision-making process – at the border between the informal and the formal consultations – play an important process in the *a priori* validation of the scientific tools. Although they are not sufficiently addressed in the EU decision-making literature, the various conferences, workshops, deliberative mechanisms, consultations with stakeholders, working groups or other expressions of interest help to get the different Commission services and stakeholders acquainted with the tools, with their strengths and weaknesses.

---

[49] Pierre Audiger underlines that in the Call for Tender specifications half of the conclusions are implicitly there: "Avant de prendre de nouvelles initiatives, la Commission commandite généralement des études à des consultants extérieurs; mais c'est la Commission qui rédige les cahiers des charges et chacun sait que, en la matière, 50% des conclusions sont implicites dans les cahiers des charges". P. AUDIGER, *La politique énergétique de la Commission européenne et la concurrence: un éclairage politique*, Revue de l'Énergie, No. 577, Paris, mai-juin 2007.

This phase allows the different Commission DGs, Member States and stakeholders to check the model assumptions, directly to question the scientist, to discuss the costs of the technologies, to argue on the country or sectoral specificities, etc. It is a crucial phase in establishing the confidence with the tool.

More the model basics, functioning and results are disseminated to several DGs, governments, industry, trade-unions and NGO, more they are criticised but more they get legitimacy. As demonstrated in the energy-related policy case studies, at most the models are discussed, at most they will be used and quoted in Commission Staff working document, in Impact Assessments and in EU "soft law" (Communications, Green and White Papers).

The definition of research priorities (what kind of tools will be developed, with what geographical and technological coverage, with what reference scenario, etc.), the debates in both policy (Commission, Council and Parliament) and academic arena (in different disciplines and within a specific discipline) and the first draft proposals need about 4-5 years.

As illustrated in Figure VIII.4, when the initial proposals start to be formulated, the discussions are generally divided into two main groups: an assessment of the economic and technical potential mostly made by the "experts"; and an assessment of the social and political relevance mostly made by "policy-makers". In the middle – as a circus points between the scientific and the policy-discussion – is the "stakeholder validation" (through conferences, workshops, position papers, etc.).

Due to its institutional specificities, the Commission has a very special peculiarity. As technical and political in the same time, its presence is ensured on both side of the figure, with both experts and policy-makers. In addition, composed of various sectoral DGs, the Commission is also well placed to achieve consensus among stakeholders.

Therefore, in order to go forward an initiative needs a double consensus: from scientists (in the first group) and from policy-makers (in the second group) with the stakeholder validation as a meeting point of techno-economic and socio-political issues.

Overall, from the development of the tool (model) to the final decisions (directive or regulation) passing through the scientific and political debates largely framed through "soft law" acts (Green Papers or Communications), a period of about ten years is needed.

Figure VIII.4: Interactions and timing from initial research to policy-making

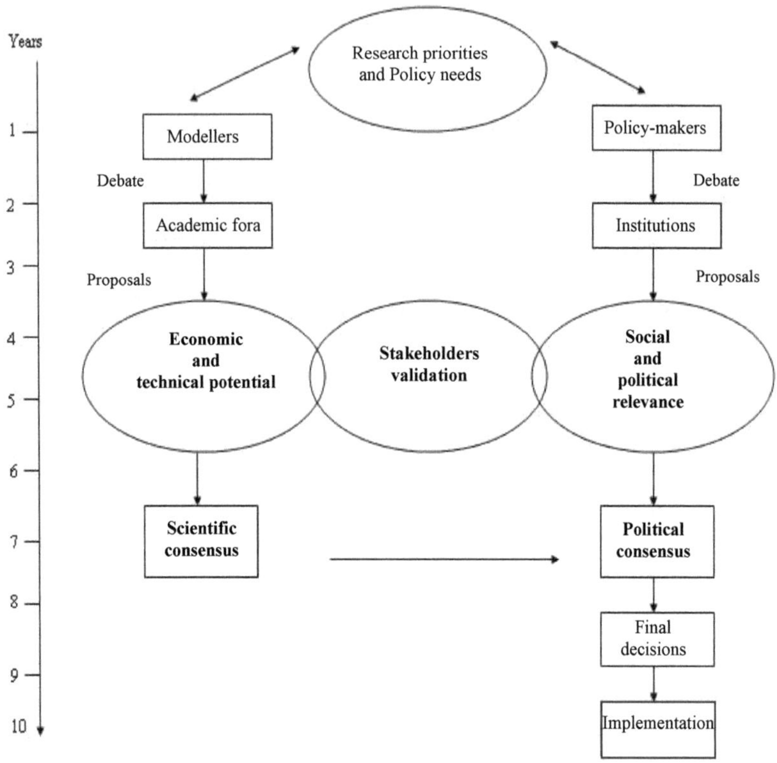

CHAPTER IX

Final Discussion and Key Messages

Summary

The final discussion is structured around six main pillars that have been detailed in the previous chapters and form which key messages have been underlined.

First of all, although the energy and research fields are at the origin of the EU integration (since the 1950s), today, energy and research policies are only partly "communautarized". But they are becoming strategic both in terms of policy relevance and in EU budget. It also worthwhile mentioning that energy economic research, and in particular modelling, has a long tradition in the EU.

Secondly, more uncertainties increase more the role of scientists in decision-making become prominent. In the technical policies, like energy and environment policies, there is an increasing interest in both qualitative and quantitative analyses. If the qualitative foresight covers well the complexity of the society, the quantitative models have a better operational policy use.

Thirdly, the combination of economic Top-down models with the technology-rich Bottom-up models could lead to an ideal hybridization. From an historical and comparative assessment, the Economics, Energy and Environment (E3) models appear as very reliable on the anticipation of the total energy consumption but rather weak on identifying the future share of fuels and technologies in the energy mix. The questions on how to better formalize socio-economic behaviours and technological progress remain valid even after thirty years of research efforts in modelling.

Fourthly, it has been demonstrated that the Top-down models are used in the EU economic and fiscal policy while the Bottom-up models are intensively used by the Commission for the climate change policy. The quantitative methods have almost the monopole in the European sectoral policies. The quantification and the possibly internalisation of external costs is under way in the EU.

Fifthly, there is a clear interaction between modelling and EU policies, between science and decision-making at EU-level. During the whole EU policy cycle, experts are associated to the decision proposals. The administration and funding of expertise can also influence the scientific results. An attempt has also been made to assess the "productivity of research" in support of EU policies.

Finally, it has been demonstrated that the scientific support is a necessary condition for a successful Commission initiative, at least in most of the energy and environment fields, i.e. that the proposal is adopted by the Commission and later by the Parliament and the Council. *De facto*, the credibility of the Commission energy and environment policy initiatives is partly in the hand of modellers that are particularly present in the up-stream European policy process.

EU policies in the energy and research fields

Energy and research are at the origin of the EU integration but are only partly "communautarized"

Energy and research polices, although less known than European agricultural or competition policies for example, have been at the origin of the European integration and are closely linked.

The first "European Communities" in the 1950s dealt with coal and nuclear energy. It was the European Coal and Steel Community (ECSC) and the European Atomic Energy Community (Euratom)[1]. Nevertheless, energy is almost totally absent of the EU primary legislation. In the current European (Nice) Treaty, energy is quoted only once under the Trans-European Network chapter. All the EU policies in the energy field (liberalisation of the energy markets, directives on renewables, on energy efficiency, etc.) have been adopted through other legal basis and more specifically to three articles of the Treaty concerning environment, internal market and approximation of laws[2].

Research at the European level was originally launched in the field of energy. RTD in coal and nuclear (and in the non-nuclear field after the 1973 oil crisis) was already carried out at the EU-level well before

1 N. LUCAS, *Energy and the European Communities*, Europa Press, 1977.

2 In the 2004 "Constitutional Treaty" and in the 2007 Lisbon Treaty, energy becomes a "shared competence" between the Union and the Member States. The EU policy on energy focused on three items: ensuring the functioning of the energy market; ensuring security of energy supply; and, promoting energy efficiency and saving and the development of new and renewable forms of energy. Nevertheless, unanimity at the Council is still required for the measures significantly affecting a Member State's choice between different energy sources and the general structure of its energy supply. If the Lisbon Treaty enters into force, it most probable that new EU energy-related policies would be more frequent (cf. "Community method" and Commission power of initiative).

the existence of the European RTD "Framework Programme" in the mid-1980s. The EU Joint Research Centre was founded with the 1957 Euratom Treaty. Research (and technological development) was introduced in the Single European Act (1985) and became a European policy in the Maastricht Treaty (1991). Since the middle of the 1980s and the successive RTD Framework Programmes, research started to include other fields than energy. This was the case, for example, for information technologies, new materials or genomics.

Energy and research are becoming strategic policies for the EU

In 2000, two European Commission documents launched energy and research at the core of the EU policy debate: the Green Paper on energy security of supply and the Communication on the European Research Area[3].

Following the Green Paper, several European Directives have been adopted in the energy field. In January 2007, the Commission presented an Energy package including the Communication "An energy policy for Europe"[4]. The 8-9 March 2007 European Council adopted ambitious energy and environment related targets. At the informal Lisbon European Council of 18-19 October 2007, the Heads of State and Government reached an agreement on the new Lisbon Treaty in which energy, as in the 2004 Constitutional treaty, became a shared competence between the EU and the Member States. On 11-12 December 2008, the European Council succeeded in translating highly political objectives into legally binding texts[5].

Concerning research, in the follow-up of the European Research Area and the Lisbon strategy (2000) where RTD has a key role to play in the expected competitive knowledge-based society, the Barcelona European Council (2002) requested to achieve 3% of GDP devoted to RTD by 2010.

De facto, energy is becoming a key EU subject (one third of the German presidency conclusions in March 2007 concern energy vs. a few lines in the past). Research is emerging as a major EU policy in budgetary terms (almost € 54 Billion for the seventh RTD Framework Programme covering the period 2007 to 2013, i.e. the third policy after

---

3 European Commission, *Towards a European strategy for the security of energy supply*, COM(2000)769; European Commission, *Towards a European Research Area*, COM(2000)6.

4 European Commission, An energy policy for Europe, COM(2007)1.

5 On the discussions about the climate change and energy package for 2020, several difficulties of this European Council concerned the distribution of effort ("burden-sharing") between Member States and sectors and the degree to which the allowances used to limit emissions should be sold at auction.

agriculture and structural funds). Energy technologies are being more and more pushed at the European level[6].

The main objectives are to establish a genuine EU energy policy (internally and externally) and to consolidate the European Research Area with leverage on public and private RTD.

Energy economic research, in particular modelling, has a long tradition in the EU

Since 1975, the European Commission promoted the development and use of different kind of models (cf. Energy Systems Analysis and Modelling programme). This research work was, from its origin, closely linked to energy policy formulation.

With the birth of the RTD Framework Programmes in 1984, "Energy economic research" also called "Energy strategy", "Generic energy activity", "Socio-economic research" and "Knowledge for energy policy-making" was always a research theme in the successive Energy programmes since the first to the seventh RTD Framework Programme.

The subjects covered by this energy economic research make the links between energy and environment and address the issues of natural resources, economic growth and social needs, including both market competition and environmental constraints with short and long-term views. In a certain sense, EU energy economic research covered most of the issues that will be raised thirty years later in the sustainable development context where energy and environment, science and policy, technology and society are closely interlinked.

> Research into sustainable development must include short-term decision support projects and long-term visionary concepts and has to tackle problems of a global and regional nature. It has to promote inter – and transdisciplinary approaches involving social and natural sciences and bridge the gap between science, policy-making and implementation[7].

The majority of the EU energy economic research dealt with model developments (Top-down and Bottom-up), scenarios building and energy externalities quantification, i.e. with scientific tools and methods aiming to assess the impacts of energy-related policies.

The budget allocated to EU energy economic research has been continuously supported and estimated to be about € 4 Million per year since 1990. This budget has been rather constant, but small compared to the

---

6 European Commission, *A European strategic energy technology plan (SET-Plan) – Towards a low carbon future*, COM(2007)723.

7 Council of the European Union, *Review of the EU sustainable development strategy – Renewed strategy*, 9/6/2006.

total of about € 400 Million per year relatively well distributed between EU nuclear – fusion and fission – and non nuclear research – mainly renewables, energy efficiency, fuel cells and hydrogen, CO_2 capture and storage (See Chapter II)[8].

Energy has been at the core of many European modelling activities since the end of the 1970s and the beginning of the 1980s (first RTD Framework Programme). EFOM was developed for the supply side, MEDEE for the demand side and MIDAS for the integration of the energy systems (supply and demand) through energy prices. With PRIMES and POLES, the semi-partial equilibrium model entered into the European arena. The SAFIRE first and GREEN-X later focused on renewables while MURE was dealing with energy efficiency policies and measures. For analysing the impacts on the European economy, the econometric models E3ME and NEMESIS were developed as well as the computable general equilibrium model GEM-E3. The EXTERNE accounting framework was the first and most often quoted reference to quantify the social and environmental costs of electricity and transport (see Chapter V).

With the advent of the European Research Area (cf. "A common system of scientific and technical reference for policy implementation" and a "Scientific evidence basis for policy-making"), the sixth and seventh RTD Framework Programmes further push for research activities supporting policy formulation[9].

Qualitative and quantitative tools for policy analysis

Uncertainties increase the role of scientists in decision-making

Although the energy sector is subject to some inertia (long-term investments such as power plants or electricity grids, strongly established behaviours in transport, fossil fuel based economy), more and more uncertainties appear like the consequences of climate change or the technological progress. In particular, because of globalisation, world competition, intangibles, merger and acquisition, use of information and communication technologies, and – in the energy field – because the world is operating closer to limits in terms of pollution and resources, the whole economy is more and more subject to uncertainties. As Jean-Marie Chevalier wrote:

---

8 D. ROSSETTI di VALDALBERO, B. SCHMITZ, W. RALDOW, M. POIREAU, *European Union energy research*, Revue de l'Énergie, No. 576, Paris, March-April 2007.

9 European Commission, *The European Research Area: New perspectives*, COM(2007)161.

> La multiplication des incertitudes est l'une des caractéristiques majeures des systèmes économiques (et énergétiques) contemporains[10].

If economic and energy uncertainties can be clearly identified, the European Union policy-making process uncertainties have also to be underlined. At the EU-level, it is more difficult than at the national level to predict what problem will emerge and by whom and how this problem will be presented.

> Alors qu'au niveau national, un acteur ou une coalition d'acteurs, peuvent espérer verrouiller – même relativement – le processus d'inscription des problèmes sur l'agenda politique, il est beaucoup plus difficile de l'imaginer au niveau communautaire, compte tenu de la diversité des intervenants et du flou des mécanismes de négociation[11].

In this context of uncertainties, policy-makers tend more often to support their choices on the results of scientific analysis[12]. Economic, social, environmental and technological uncertainties about the future can be reduced by anticipating possibilities and integrating them into strategies. The importance of forecast is underlined by many authors in different fields. For example, José Achache underlined that without reliable forecast, all the opinions are equivalent[13]. This is one of the main reasons explaining the development of several qualitative and quantitative future outlooks at the world, European and national level, in the public policy domain or as a business strategy (see Chapter III).

Policy-makers want to find the most reliable and least contestable knowledge base. Stakeholders need the most credible data. Most of the time, scientists are the best-placed category to provide answers to both policy-makers and stakeholders at the same time.

> The scientific added value relies in the fact that it represents the best in terms of methods, accepted knowledge and quality of the authors. In addition, the scientific conclusions are more likely to be 'right' also because,

10 J.M. CHEVALIER, *Les grandes batailles de l'énergie*, Gallimard, Paris, 2004.

11 P. MULLER, *Un espace européen des politiques publiques*, in Y. MENY, P. MULLER, J.L. QUERMONNE, *Politiques publiques en Europe*, L'Harmattan, Paris, 1995.

12 This is particularly evidenced in the technical matters. At the world level, the most obvious example is climate change where the reports made by scientists – through the IPCC – have a strong political dimension. At the EU-level, the power of the Commission, the most technocratic European institution, will be illustrated several times in this book through the energy-related policy case studies.

13 J. ACHACHE, *Les sentinelles de la terre*, Hachette, Paris, 2004.

when faced with criticism, policy-makers often need to be able to demonstrate that they relied upon the best advice available[14].

An increasing interest in energy policy analysis, both qualitative and quantitative

Interest in energy policy analysis developed these last forty years as the energy system came into conflict with other strategic issues. The main constraints were felt through the oil shocks of the 1970s (security of supply and natural resources issue), the concerns about air pollution (acid rain), the question of social and environmental damages (external costs), the global warming problem (climate change) and the "green paradigm shift" in more advanced countries (precaution and polluter-pays principles)[15].

The complex relation between energy use and GDP growth also called for more energy policy analysis. As Jean-Marie Chevalier said,

> Energy consumption must be seen in its relation to economic growth and economic development. In less than a century, commercial energy has become the engine of economic activity and, in our energy final consumption, electricity is now considered as an essential product[16].

For looking at the future energy supply and demand, for measuring greenhouse gas emission trends and for assessing impacts of alternative policies and measures, both qualitative and quantitative scientific methods have been developed and numerous studies have been published by public authorities and by private companies (see Chapters III, IV and V).

Discussion about the likely problems of the future does not lead to consensus about the policies needed to respond to those problems (taxation, emission trading scheme, targets) but it ensures that the political debate has to take a long view. As Peter Schwartz said:

> Les scénarios son un moyen d'articuler et d'ordonner la gamme des incertitudes essentielles susceptibles d'affecter les résultats des plans en cours (…). Les scénarios sensibilisent les décideurs aux changements fâcheux (…). Une fois un ensemble de scénarios différents mis au point (…), les

14 P. COLLINS, U. LINDBERG, *Independent science advice in the European Union*, IPTS Report, No. 70, December 2002.

15 R. KEMP, *Technology and transition to environmental sustainability: the problem of technological regime shifts*, Futures, Vol. 26, 1994.

16 J.M. CHEVALIER, *Energy economics and energy econometrics*, in J.H. KEPPLER, R. BOURBONNAIS, J. GIROD (eds.), *The econometrics of energy systems*, Palgrave Macmillan, New York, 2007.

cadres d'une société partagent désormais un langage nouveau en matière de planification (...)[17].

Qualitative and quantitative energy outlooks aim to provide a coherent picture of the future. They both help to make previsions and to assess the impacts of certain measures. They both allow defining a strategy that takes into account the costs and benefits of such or such a policy.

The main qualitative tools – based on expert views – are Foresight (Scenario method) and Delphi survey. Top-down and Bottom-up mathematical models are the most often used quantitative tools. For longer-term issues (more than 50 years), a mix of qualitative and quantitative analysis can be found in the Shell and IPCC scenarios as well as in the Back-casting approach[18].

Through Foresight and Models, scientists try to look at the most probable future and to limit the risks associated to such a future. In a certain sense, qualitative and quantitative outlooks aim to face uncertainty by anticipating some potential impacts coming from economic, social, environmental and technological developments.

Particularly in the energy field where most decisions have long time duration (cf. building sector, power generation and energy-intensive manufacturing), the policies of today will affect the lives of the next and subsequent generations. This is also a reason why the energy sector has a so strong tradition in Scenario method and System analysis.

Foresight covers well the complexity of the society while forecasting (models) have a better operational policy use

Qualitative and quantitative tools have both their strengths and weaknesses. The first ones have the advantage of being participatory and of addressing the whole complexity of the society (collective reflection). The quantitative tools benefits are their sectoral precision and their operational policy implementation ("what if" in a specific case and in a specific moment).

If the Scenario method has the advantage of being action-oriented and open to alternative futures (cf. "thinking the future, debating the future and shaping the future", "qualitative outlooks"), Models provide an holistic evaluation – tackling in the same time the economic, social,

[17] P. SCHWARTZ, *La planification stratégique par scénarios*, Futuribles, Paris, mai 1993.

[18] Shell, *Energy needs, choice and possibilities: scenarios to 2050*, London, 2001. IPCC, *Climate change, fourth assessment report*, Summary for policymakers, WMO/UNEP/IPCC, 2007. B. CHATEAU, T. HAMACHER, K. SMEKENS, D. MARTINSEN, H. REISINGER, M. PATEL, *Very long term energy modelling (VLEEM)*, Final report, European Commission, Brussels, 2000.

environmental and technological issues – including the costs and benefits of a specific policy or measure. As indicated by Paul Zagamé, "l'apport des modèles peut être très important par la cohérence d'ensemble dans laquelle ils permettent d'évaluer des politiques"[19].

As already pointed out by Sergio Koreisha and Robert Stobaugh at the end of the 1970s, "The world of human institutions is a world of power and politics, one not easily captured in mathematical models"[20]. Today, although models have achieved in many cases a fair degree of technological details, progresses can still be made to replicate with reasonable accuracy the behaviour of the actors and the functioning of energy markets.

The establishment of an energy outlook require deep research and a capacity of synthesis between several disciplines (e.g. between geologists and economists on oil reserves), interests (e.g. between nuclear promoters and supporters of renewables) and policies (e.g. between "competitiveness" and "sustainable development").

As Pierre Valette said, a successful foresight exercise, i.e. that covers the whole problematic and that could be used in policy-making, should have two components: expert-based contrasted socio-economic and policy scenarios (qualitative part) and a mathematical quantification of the impacts of the alternative scenarios (quantitative part)[21].

Several challenges should be faced by qualitative and quantitative methods. Both approaches have a risk of manipulation for example by the choice of experts in the qualitative method or by the technological development and substitution envisaged in the quantitative tools. At least two major issues should be highlighted:

- When dealing with Scenarios (qualitative approach), participants representing a large set of different functions and disciplines must be involved. Rigorous analysis, ongoing conversations and multiple iterations of the scenarios themselves are needed to establish the right energy outlooks. According to some authors like Kees Van Der Heijden, a "remarkable person" – i.e. someone that has sufficient courage and expertise to go beyond the conventional thinking (Business as Usual) – is required[22]. As Pierre Wack used to say,

19 P. ZAGAME, *L'expérience française de modélisation macro-économétrique: bilan et perspectives*, Revue d'économie politique, No. 5, Paris, 1987.

20 S. KOREISHA, R. STOBAUGH, *Limits to models*, in Energy Future, Report of the Energy project at the Harvard Business School, Random House, New York, 1979.

21 P. VALETTE, *European Commission*, DG Research, EU 2025, internal note, 21/11/2007.

22 K. VAN DER HEIJDEN, R. BRADFIELD, G. BURT, G. CAIRNS, *The Sixth Sense*, Wiley, New York, 2002.

"remarkable people" are those with unconventional insights[23]. In the theory of Jacques Durand, this would allow to go from a "Continuity future" to an "Exploratory" or even to a "Breakthrough" future[24].

- When dealing with Models (quantitative approach), technological potential should not be over-estimated by learning curves (decreasing cost curves), especially in relation to social demands of a future energy system[25]. For example, PRIMES, POLES and GEM-E3 take into account technical progress induced by the changes in the economy, after the additional energy charges or the substitutions in production factors. They allow looking at the opportunities to make low-cost energy savings, where investments could be rapidly repaid (shift towards better technology including premature replacement of equipment, changes in the fuel mix, etc.)[26]. On the contrary, institutional changes and social behaviour patterns are too often marginally considered although they are crucial in shaping the future energy system[27].

Economics, energy and environment (E3) modelling

The combination of economic Top-down models with the technology-rich Bottom-up models could lead to an ideal hybridization

Top-down models are strong in terms of economic explicitness (micro-economic realism and macro-economic completeness) and Bottom-up models are technology-rich. Top-down models are particularly useful for studying the interactions between the energy system and

---

23 P. WACK, *Scenarios: Uncharted waters ahead*, Harvard Business Review, Sept-Oct. 1985.

24 J. DURAND, *Prospective, discontinuité et instabilité*, Futuribles, Paris, Automne 1975.

25 M. GRUBB, J.E. EDMONDS, *The cost of limiting fossil fuel CO_2 emissions – a survey and analysis*, Annual Review of Energy and Environment, Vol. 18, 1997.

26 P. CAPROS, *Energy technology dynamics and advanced energy systems modelling*, International Journal of Global Energy Issues, United Kingdom, 2000. P. CAPROS, L. MANTZOS, P. CRIQUI, N. KOUVARITAKIS, A. SORIA RAMIREZ, L. SCHRATTENHOLZER, E.L. VOUYOUKAS, Climate technology strategies, Physica-Verlag, Berlin, 1999.

27 J.M. MARTIN, *Le changement des technologies de l'énergie: genèse, modalités et hypothèses explicatives*, in B. BOURGEOIS, D. FINON, J.M. MARTIN, *Énergie et changement technologique – une approche évolutionniste*, Economica, Paris, 2000. B. CHATEAU, B. LAPILLONNE, *Energy demand: facts and trends – a comparative analysis of industrialized countries*, Springer-Verlag, Vienna, 1982. B. LAPONCHE, B. JAMET, M. COLOMBIER, S. ATTALI, *Maîtrise de l'énergie pour un monde vivable*, ICE, Paris, 1997.

the rest of the economy whether Bottom-up models describe energy technologies in details from both supply and demand-sides.

Top-down models describe the energy system in terms of aggregate relationships derived empirically from historical data. Bottom-up models determine the financially cheapest way to achieve a specific energy or environment target based on the best available technologies and processes. As noted by Jean-Charles Hourcade, Mark Jaccard, Chris Bataille and Frédéric Ghersi, Top-down models represent technological change as an abstract, aggregate phenomenon – implicit in their substitution elasticities. This approach is well suited to helping policy-makers assess economy-wide price instruments such as taxes and tradable permits, but has difficulties in assessing the combined effect of these price-based policies with technology-specific policies[28].

The choice of specific technologies under alternative scenarios is of interest to policy makers, who tend to avoid policies not easily accepted by society like taxation. Policy-makers often prefer softer or technology-oriented instruments like subsidies, green certificates or labelling programs. In other words, Top-down models have considerable difficulty addressing the type of questions that interest policy makers. Already twenty years ago, Paul Zagamé underlined the fact that the macro-economic exercises were too much aggregated to answer to specific and sectoral questions, "Les prévisions et les exercices variantiels macro-économiques étaient trop agrégés pour répondre au besoin d'information des partenaires sociaux (entreprises, syndicat, certaines administrations)".

The major studies led by the EU, the IEA or the US DoE demonstrate that Bottom-up models (POLES, WEM, NEMS/SAGE) are often preferred for evaluating the proposed measures in the energy sector (see Chapter VII)[29]. In addition, as Dominique Finon said, Bottom-up models have achieved a good level of details, sufficient preciseness and transparency of assumptions[30].

---

28 J.C. HOURCADE, M. JACCARD, C. BATAILLE, F. GHERSI, *Hybrid modelling: New answers to old challenges*, The Energy Journal, Special issue, IAEE, Cleveland, 2006.

29 European Commission, *World Energy Technology Outlook – 2050 (WETO-H2)*, Luxembourg, 2006; International Energy Agency, *World Energy Outlook (WEO)*, Paris, 2006; US Department of Energy (EIA), *International Energy Outlook (IEO)*, Washington, 2006.

30 D. FINON, *Prospective énergétique et modélisation – Identification de pistes de progression méthodologique*, Note au Conseil Scientifique de l'Institut Français de l'Énergie, Paris, Juin 2003.

Jean-Charles Hourcade *et al.* highlighted the different options that Top-down and Bottom-up modellers can pursue[31]:

- A Top-down model that partly renounces the macroeconomist's toolkit – constant elasticities of substitution and the autonomous energy efficiency index – and relies on innovative ways to represent not only energy supply but also energy end-use technologies as described by Bottom-up analysis, and technology adoption as described by microeconomic studies, especially regarding households.
- A Top-down model that increases its disaggregation level to include a reduced-form Bottom-up module of some part of the energy system (e.g. in energy supply or the transport sector).
- A Bottom-up model that includes empirically estimated microeconomic parameters related to technology choice; functions to clear markets for energy, other intermediate inputs, and final goods and services based on changes in the cost of production, using either price elasticities or more advanced Computable General Equilibrium techniques (consumer utility; firm profit functions; functions to balance government budgets, exchange rates, capital and labour markets).
- A composite hybrid model that includes all of the major theoretical and structural characteristics of the most advanced Top-down models along with the major characteristics of the most advanced Bottom-up models, with technological detail in all sectors and behavioural parameters that are empirically estimated from microeconomic and macro-economic research. While such a model would present the greatest challenge in terms of theoretical consistency, mathematical complexity and empirical estimation, it nonetheless represents an objective that some modellers might aspire to, and has been colloquially referred to as the "Holy Grail".

According to J.C. Hourcade *et al.*, beyond their fruitful direct application to policy analysis, hybrid models should ultimately be regarded as a major avenue of progress in the modelling discipline. They are communication tools between various fields of knowledge (engineering, macro- and micro-economic analysis, comparative institutional studies), and one of their most important contributions may be to help detect missing information and dynamics, and to provide a structure for discussion and progressive consolidation of modelling methods.

[31] J.C. HOURCADE *et al.*, *Hybrid modelling: New answers to old challenges*, *op. cit.*

Reliability of models on total energy consumption but weaknesses on individual fuel share and on technological progress

The *ex-post* evaluation of model results from 1985 demonstrated that the projections for 2000 were concomitant with the real data concerning the total energy consumption for the ten Member States of the European Community of that time (see Chapter VII)[32].

Between the results of the two models used (energy demand forecasting model MEDEE and supply techno-economic energy model EFOM) and the real statistics coming from Eurostat, the deviation was of less than 2%. Nevertheless, relatively large differences have been noticed between projections and statistics concerning individual fuel and technology. For example, the models sub-estimated by 30% the natural gas consumption for the ten European Community Member States.

In agreement with Paul Zagamé ("La précision des prévisions passées rejaillit sur l'audience des prévisions d'aujourd'hui"), one can conclude that in terms of projections, modelling tools are very reliable for the total forecasted energy consumption but provide only partially good results in terms of individual fuel and technology[33].

Rather than weaknesses of the EFOM and MEDEE models assumptions and structure, these large differences are essentially due to:

- Unanticipated strong political decisions (environmental regulation, closing of mines, liberalisation of energy markets, permission to use natural gas for base-load power generation, mechanisms to support renewable energy sources),
- Social behavioural changes (cf. strong increase in the transport demand) and
- Modification of the definition and availability of statistical data (cf. renewables statistics)[34].

---

[32] J.F. GUILMOT, D. McGLUE, P. VALETTE, C. WAETERLOOS, *Energy 2000*, European Commission, Luxembourg, 1986; B. LAPILLONNE, B. CHATEAU, *The MEDEE models for long-term energy demand forecasting*, International Conference of Dublin, *Energy systems analysis*, 9-11/10/1979, D. Reidel-Kluwer, ECSC-EEC-EAEC, Brussels and Luxemburg, 1980; D. FINON, *Scope and limitations of formalized optimization of a national energy system*, in A. STRUB (ed.), *Energy models for the European Community, op. cit.*

[33] P. ZAGAME, *L'expérience française de modélisation macro-économétrique: bilan et perspectives*, Revue d'économie politique, No. 5, Paris, 1987.

[34] P.A. PILAVACHI, Th. DALAMAGA, D. ROSSETTI di VALDALBERO, J.-F. GUILMOT, *Ex-post evaluation of European energy models*, Energy Policy, Vol. 36, 2008.

The *comparison* of the assumptions and results of three different Bottom-up model-based studies led by the European Commission (EC-WETO), the International Energy Agency (IEA-WEO) and the US Department of Energy (US-IEO) provides a sort of "benchmark" and allows to highlight where the main differences are expected by 2030.

Having similar assumptions (population and GDP), the comparison indicates an overall high concordance on the projections of the "Reference case" on total energy consumption and carbon dioxide emissions from the three world energy studies. However, relatively important differences are noticed for individual energy sources and technologies, especially regarding coal, gas and nuclear. This is mainly driven by the different fuel mix used for power generation.

The US-IEO study forecasts significantly higher coal consumption than the other studies. In the EC-WETO study, the lower consumption of coal is compensated by natural gas until 2020, and by nuclear energy between 2020 and 2030. Oil accounts for about 33% in the total energy consumption in the three studies by 2030. Renewable electricity – including hydro and high growth rates for wind and solar technologies – accounts for about 20% of total power generation in the three studies by 2030.

Thirty years of research efforts in modelling but the same questions remain valid

As Henry Wan wrote in 1979

> Economics is for prediction. But even with full information and in the absence of randomness, can one always predict a unique outcome? My answer is no. Uniqueness can be guaranteed only by adopting very strong assumptions on technology and individual behaviour[35].

On the same year, John Weyant highlighted that the results from quantitative models were cited more frequently at the US Congress in the crude oil and natural gas pricing debates (1974-1978) than ever before.

> Whether the proliferation of analytic studies led to better decisions is of course very debatable. Many legislators undoubtedly quoted results from analytical studies without understanding or evaluating the assumptions that went into them. These individuals had obviously already decided how to vote on the issues based on moral, philosophical, or ethical considerations[36].

---

35 H.Y. WAN, *Causally indeterminate models via multi-valued differential equations, Bifurcation theory and applications in scientific disciplines*, Annals of New York Academy of Sciences, Vol. 316, 1979.

36 J. WEYANT, *The role of models in the oil and gas pricing debates*, Energy Modelling Forum, Stanford University, April 1979.

There are few areas of human activity in which there has been as much modelling of long-term trends as in the energy sector as Jacques Lesourne underlined at the end of the 1970s.

> The main reason for energy modelling is the huge problem posed by the transition to the post-oil era. (…). I can discern two more specific causes. The first is the small number of primary energy sources – sources which are clearly identifiable and largely interchangeable. Hence the possibility of choices, but choices which are at once too complex to be studied without any formalization and sufficiently simple to lead to models of reasonable size. The second is the slow-moving nature of energy phenomena. Every specialist knows that it takes decades for the pattern of a developed country's supply-demand balance for primary energy to change substantially. Thus the decisions that are taken commit the future over long periods. Unlike many other sectors where short-sightedness can be defended on grounds of uncertainty, energy is an area where it is essential that the long-term is taken into account[37].

In 1979, speaking about the interdisciplinarity of energy policy and its various related concerns (environmental values, nuclear safety, etc.), Alan Manne, Richard Richels and John Weyant wrote

> Models can yield only partial insights (…). One can hope that processes such as counter modelling and third party assessment will help to distinguish between objective analysis and one-side advocacy (…). The modelling community is just beginning to address the broader social and political aspects – health and environmental effects, safety considerations, and nuclear proliferation – of alternative policies[38].

The same authors also pointed out thirty years ago that "market imperfections – cartels, regulation, price controls – are the rule rather than the exception in today's energy markets. Existing energy policy models, on the other hand, rely predominately on the theory of perfectly competitive markets".

E3 modelling in EU decision-making

The use of Top-down models in EU economic and fiscal policy

The macro-econometric models have been extensively used in Europe. HERMES served the Commission to evaluate the consequences of the completion of the European Internal Market at the end of the 1980s and, associated with MIDAS, to quantify the "double dividend"

37 J. LESOURNE, *Methodological problems in long-term energy modelling*, International Conference of Dublin, *Energy systems analysis*, 9-11/10/1979, D. Reidel-Kluwer, ECSC-EEC-EAEC, Brussels and Luxemburg, 1980.

38 A. MANNE, R. RICHELS, J. WEYANT, *Energy policy modelling: a survey*, Operations Research, Vol. 27, No. 1, 1979.

(positive for the environment and positive for the employment). These figures were incorporated in the 1993 Jacques Delors White Paper on growth, competitiveness and employment[39].

GEM-E3, E3ME and HERMES were used for the Community framework for energy taxation[40]. In 1997, they assessed the macro-economic impacts of the proposed Directive on the taxation of energy products. The models concluded that when recycled to reduce the social security contributions the proposed increase in the EU minimum rate on motor fuels, heating fuels and electricity would have very marginal – and positive – impacts on GDP (from 0.06% to 0.2%) and employment (from 155,000 to 457,000 jobs)[41].

In the revised proposal, that became the 2003 Directive to restructure the Community framework for the taxation of energy products and electricity, only the GEM-E3 model was used[42]. It concluded that approximately no change in GDP and employment would occur from the Community framework for energy taxation (respectively from -0.01% to 0.00% for GDP and from -0.01% to 0.02% for employment).

If the modelling results were not sufficient to argue in favour of an ambitious minimum level of taxation (the final 2003 Council Directive was very modest in comparison to the 1997 Commission Proposal), these modelling studies helped in achieving a consensus of all the Member States (unanimity) at the Council.

Later, macro-econometric models have continued to be used to assess EU policies. For example, in 2005, NEMESIS was the basis of the analysis of the impact of the EU objective to achieve a 3% of GDP devoted to Research[43].

The intensive use of Bottom-up models in EU climate change policy

The Bottom-up POLES and PRIMES models served as analysis tool including the relationship between energy policy and technology as-

---

39 European Commission, *White Paper on growth, competitiveness and employment: the challenges and ways forward into the 21st century*, COM(93)700.

40 D. ROSSETTI di VALDALBERO, *Energy market regulation: Impacts of EU research*, in Ch. BÖHRINGER, A. LANGE, *Applied research in environmental economics*, ZEW-Physica-Verlag, Heidelberg, 2005.

41 European Commission, *Proposal for a Directive restructuring the Community framework for the taxation of energy products*, COM(97)30.

42 Council *Directive 2003/96/EC restructuring the Community framework for the taxation of energy products and electricity*, OJ, L 283, 27/10/2003.

43 European Commission, Staff working document, *Annex to the proposal for the Council and European Parliament decisions on the seventh EU RTD Framework Programmes (EC and Euratom)*, COM(2005)119 and SEC(2005)430.

sessment. They focus on market-mechanisms influencing the evolution of energy demand and supply and on the technology penetration in the market. The need to represent the growing process of market liberalisation motivated researchers to adopt market-oriented modelling approaches giving rise to these "new generation models".

POLES and PRIMES have been key scientific references in the EU climate change policies. They were used for preparing the negotiation position of the EU for the Kyoto conference on climate change. PRIMES in particular was used to prepare the EU energy and emissions outlook for the Shared Analysis project between the European Commission and the Member States[44].

PRIMES gave the main EU reference for the future increase in CO_2 emissions where the "Baseline" expected an increase of 8% of CO_2 emissions by 2010 in relation to 1990. This Baseline will be used in order to calculate the real effort of Kyoto, also considering the distribution of efforts among sectors. PRIMES provided the initial estimates of the "burden-sharing" among EU Member States[45].

POLES and PRIMES have been used to compute the least-cost option to achieve a certain greenhouse gas emission reduction target. They simulated CO_2 emission constraints for the Kyoto first commitment period (2008-2012) and for longer-term EU and international climate regimes (2020, 2030, 2050). They produced insights on abatement costs by countries, regions and sectors[46].

POLES and PRIMES (associated to GEM-E3) provided the economic analyses of the impacts of Kyoto flexibility instruments. In particular, they were the main basis for evaluating the EU-wide Emission Trading Scheme (ETS) – and by this way for giving the economic justification of such innovative policy in the Commission Green Paper ("emission trading could reduce the cost of Kyoto by up to 30%"). They gave the costs and benefits of trading, their allocation among sector and countries[47].

The ETS Green Paper will be followed by a proposal for a Directive and then a Directive – adopted in 2003 – establishing a scheme for greenhouse gas emission allowance trading within the Community.

44 European Commission, *Communication on climate change – The EU approach for Kyoto*, COM(97)481.

45 European Commission, *Preparing for implementation of the Kyoto Protocol*, COM(99)230.

46 European Commission, *Communication on climate change – Towards a EU post-Kyoto strategy*, COM(98)353.

47 European Commission, *Green Paper on greenhouse gas emissions trading within the European Union*, COM(2000)87.

POLES and PRIMES were also the basis of the extended Impact Assessment on the "linking Directive"[48]. Modelling results were extremely useful in preparing the new EU position in terms of Climate Change policy, i.e. very favourable to the Kyoto flexible instruments, when it was not at all a EU position before Kyoto.

After 2005, the further policy application of Bottom-up models can be seen, for example, in the EU targets for reducing greenhouse gas emissions (20% by 2020) that were established following the POLES simulations[49].

Bottom-up models – like NEMS/SAGE or WEM – are also intensively used by other organisations like the Department of Energy (EIA) in the USA or the International Energy Agency for their *International Energy Outlook* and *World Energy Outlook*.

The almost monopolistic use of quantitative tools for specific EU sectorial policies

The computer based framework SAFIRE provided the basis for setting the EU objective to double – from 6 to 12% – the share of renewables in the gross inland energy consumption by 2010[50].

Three years later, in the proposal for a Directive, SAFIRE provided the main figure on EU and national targets for the share of electricity produced from renewable energy sources[51]. This became law in the Directive on the promotion of renewable energy sources electricity, that aimed to increase green electricity from 14% in 1997 to 22% by 2010[52].

---

48 POLES and PRIMES provided the detailed world and EU energy scenarios with outlooks on energy demand, supply, trade and prices by world region and Member States. These scenarios will be used in several EU policy applications, including for fixing the European Union 2007 "Energy and climate change package" targets (See Chapter VI). In addition, technology scenarios were studied either based on exogenous hypotheses defining possible major technological breakthroughs, either based on an endogenous technological change approach that takes into account the impact of learning by doing and of R&D on technology performance. POLES was also largely used in the research field through the *World Energy Technology Outlook* study (WETO).

49 European Commission, *Limiting global climate change to 2 degrees Celsius – The way ahead for 2020 and beyond*, COM(2007)2.

50 European Commission, *Energy for the future: Renewable sources of energy*, COM(97)599.

51 European Commission, *Proposal for a Directive on the promotion of electricity produced from renewable energy sources in the internal electricity market*, COM(2000)279.

52 European Parliament and Council *Directive 2001/77/EC on the promotion of electricity produced from renewable energy sources in the internal electricity market*, OJ, L 283, 27/10/2001.

For the further EU policy applications related to renewables (after 2005), including the Impact Assessment of the 20% renewable target by 2020, the GREEN-X tool was used[53].

In relation to energy efficiency policies in the EU, the electronic database MURE provided the basis for the main target of energy efficiency in EU, i.e. 18% improvement by 2010 compared to the baseline. This figure was published in the Commission energy efficiency Communication and Action plan[54].

In the proposal for a Directive on the energy performance of buildings, the saving potential of 22% (heating, hot water, air-conditioning or lighting purposes) in buildings also came from the MURE database[55]. In the proposal for a Directive and in the Directive on energy end-use efficiency, MURE (merged with the ODYSSEE database) provided the energy efficiency potential in Europe[56].

The 2005 Green Paper on energy efficiency stated that the EU could save at least 20% of its present energy consumption in a cost-effective manner, equivalent to € 60 Billion per year. This statement was derived from several scientific studies including those performed with MURE and ODYSSEE[57]. The Spring 2007 European Council adopted the 20% energy efficiency improvement targets (in relation to the baseline)[58].

The quantification of external costs is possible and its internalisation in the EU is under way

The measurement of social and environmental damages, i.e. of external costs, was considered almost impossible until the end of the 1980s. Thanks to large EU research efforts (€ 10 Million in 10 years), the EXTERNE methodology was developed and adopted by the majority of scientists in the field.

---

53 European Commission, *Renewable energy road map – Renewable energies in the 21st century: building a more sustainable future*, COM(2006)848.

54 European Commission, *Energy efficiency in the European Community – towards a strategy for the rational use of energy*, COM(98)246; European Commission, *Action plan to improve energy efficiency in the European Community*, COM(2000)247.

55 European Commission, *Proposal for a Directive on the energy performance of buildings*, COM(2001)226.

56 European Commission, *Proposal for a Directive on energy end-use efficiency and energy services*, COM(2003)739; European Parliament and Council *Directive 2006/32/EC on energy end-use efficiency and energy services and repealing Directive 93/76/EEC*, OJ, L 114, 27/4/2006.

57 European Commission, *Green Paper on energy efficiency or doing more with less*, COM(2005)265.

58 European Commission, *Action Plan for energy efficiency: Realising the potential*, COM(2006)545.

EXTERNE became a major world reference and its results have been widely used in the EU policies. In the environmental field, EXTERNE quantified the benefits of limiting certain pollutants into the air from large combustion plants. Close to ECU 4 Billion of benefits per year were mentioned[59].

EXTERNE was the main tool for the cost-benefit analysis of the Community strategy on air pollution (CAFÉ) published in 2005. The figures of 60,000 avoided premature deaths and between € 42 and 135 Billion of health benefit per year in 2020 coming from the EU Strategy were quoted in the Commission EXTERNE-based Impact Assessment[60].

In the energy and transport fields, EXTERNE was often mentioned and was the main basis for the Commission to quote that "Fiscal instrument should lead to the internalisation of damage caused to the environment"[61]. On year later, the Commission stated that the EU should be heading "Towards modal rebalance and greater internalisation of external costs which may be achieved by replacing existing transport system taxes with more effective instruments for integrating infrastructure and external costs"[62]. Concerning the charging of heavy goods vehicles for the use of certain infrastructures, it was recognised that "a number of studies use methods that enable monetary values to be attributed to externalities"[63].

EXTERNE was the major reference in the Community guidelines on State aid for environmental protection where the renewable energy producer can receive up to € 0.05 per kWh due to their "avoided external costs"[64].

---

59 European Commission, *Proposal for a Directive amending Directive 88/609/EEC on the limitation of emissions of certain pollutants into the air from large combustion plants*, COM(98)415.

60 European Commission, *Thematic strategy on air pollution*, COM(2005)446.

61 European Commission, *Towards a European strategy for the security of energy supply*, COM(2000)769.

62 European Commission, *White Paper on European transport policy for 2010*, COM(2001)370.

63 European Commission, *Proposal for a Directive amending Directive 99/62/EC ("Eurovignette") on the charging of heavy goods vehicles for the use of certain infrastructures*, COM(2003)448.

64 European Commission, *Community guidelines on state aid for environmental protection*, OJ, C 37, 3/2/2001.

Modelling and EU policies interaction and efficiency

Administration and funding of expertise can influence scientific results

The Commission has various ways of collecting expertise. Sometime scientists and interested parties are brought together in expert groups or the Commission may set up workshops, hearings or conferences. Expertise may also be gathered by the Commission publishing a Green or a White Paper. The Commission may also call upon the Joint Research Centre (JRC), set up specialised agencies or entrust studies to external academics and consultants.

> Increasingly, the interplay between policy-makers, experts, interested parties and the public is part of policy-making, and attention needs to be focused not just on *policy outcome* but also on the *process* followed (…). Expertise is involved at all stages in the policy cycle, embracing the initial identification of the need for a policy action (including foresight exercises to identify future trends), shaping policy options (including Impact Assessment), policy proposal, policy implementation as well as policy monitoring and review[65].

The models – like EFOM MEDEE, GEM-E3, NEMESIS, POLES, PRIMES, SAFIRE or MURE often quoted and used in EU policy-making between 1995 and 2005 – were initially developed under funding from the Commission DG Research. These models were firstly funded through the Call for Proposals procedure where the projects are evaluated by scientists (peer-reviewers). Only in a second step, through Call for Tenders, the models are employed for precise analytical exercises requested by the policy DGs (like DG Energy and Transport, DG Environment or DG Taxation).

Most of the studies in support to policies (Communications, Proposals for Directives, etc.) are issued at the last moment following the most recent policy debates, i.e. only some days before the formal adoption of the official document. Concretely, the specific tendered studies (for policy DGs) did not get the time to be scientifically peer-reviewed. In fact, while a key objective of DG Research is the development of knowledge in Europe, most of the policy DGs are hampered by the short-term ("the proposal for a Directive should be presented by the Commissioner in the next four months" could often say a Director General to its staff).

Therefore, EU scientists are "purely researchers" in the Call for Proposals phase (with little interference from policy-makers) while they are

---

[65] European Commission, *Communication on the collection and use of expertise by the Commission: Principles and guidelines – Improving the knowledge base for better policies*, COM(2002)713.

"almost consultants" in the Call for Tenders phase (under a clear policy pressure). If the objectivity of the figures provided by the scientists in the initial model runs is not often raised into question in the first methodological phase, it becomes more sensitive in the last policy phase.

The distinction between Call for proposals (almost complete freedom of scientists) and Call for Tenders (specific runs for specific policy purposes), makes important – for the credibility of the Commission initiative – that the models are scientifically agreed and validated by peer-review before the Call for Tender[66].

Productivity of EU policy-oriented research and new indicators

This work allowed to attempt to measure the productivity of research in support to EU-policy-making. The productivity is here defined as the ratio between the EU energy economic research expenditure *versus* the use of the model in specific policies.

The development, updating and use of the European models and tools (GEM-E3, POLES, PRIMES, SAFIRE, MURE and EXTERNE) required about € 30 Million in ten years. This cost has to be compared to their applications in five major EU policy areas between 1995 band 2005: energy taxation, climate change policy including emissions trading within Europe; renewable energy, energy efficiency and internalisation of external costs.

The development of models has efficiently contributed to the support of EU policies. In comparison to the impact of the policies presented by the Commission, the costs dedicated to the development of these tools seem proportional to the objectives to be achieved. For example, the external costs of electricity and transport have been evaluated by the

[66] The belief that economic analysis will lead to less regulation ("Paralysis by analysis"), for example due to the difficulty to demonstrate the benefits of the proposed regulation or to the lack of scientific data and consensus on a certain issue, has not been demonstrated in the US where economic analysis can lead to stronger as well as weaker regulation (R. MORGENSTERN, *Economic analysis at EPA: Assessing regulatory impacts*, Resources for the Future, Washington, DC, 1997). In Europe, following the five case studies (Chapter VI), one can conclude that the publication by the Commission of a Call for Proposals and then of a Call for Tender on a specific topic is already a sign of interest towards further legislation on this topic. Due to the specificities of the EU political system (shared competence between the EU and the Member States and Commission initiative power), the question is often not between "strong or weaker regulation" but between "having or not having" a EU regulation. This was clear for example regarding the TERES studies in the renewable energy field in the middle of the nineties that will give birth to the 1997 White Paper on renewables and the further EU initiatives aiming to increase the share of renewables in the EU energy mix.

EXTERNE research teams at € 120 Billion per year in the European Union[67].

Without quantified background modelling information, how Member States could have been convinced about the costs of EU energy taxation? What would have been the basis for fixing the EU Kyoto targets, firstly internationally (Kyoto) and later internally ("Burden-sharing")? What would have been the economic justification for creating an Emission Trading Scheme in Europe? How a renewable or energy efficiency objective could have been fixed? How the external costs could have become a leitmotiv in EU energy-related policies without a generally accepted scientific methodology across Europe?

A "Policy citation index" – that does not currently exist – could be a useful indicator to record the number of scientific (modelling) results that directly feds EU policy formulation. This Policy citation index could be elaborated through a monitoring of Commission up-stream policy documents (Communications, Green and White papers, and Directive proposals). It would allow looking at the ratio of money spent on tendered studies top money spent on research projects.

This would be useful for two reasons: for checking the value for money of European research in support to policy-making; and for a democratic transparency reason: to provide the Commission scientific background information to elected policy-makers (European parliament) and to representative of the Member States (Council).

In terms of research planning, from the initial development of the tool (model) to the implementation of the new scientifically supported policy ten years are needed trough a long process of academic, political and stakeholder validation where the Commission plays a key role as a "techno-political institution".

Both DG Research and the policy DGs (DG Environment, DG Energy and Transport, DG Taxation) should be involved in the modelling exercise. The first one to ensure the scientific basis and peer-review process and the second ones to specify the policy specificities in the tendering process. Inter-DG collaboration is a key success factor for modelling results implementation into policies.

---

67 D. ROSSETTI di VALDALBERO, *Quantification and tentative internalisation of energy external costs in the EU*, in SPITZER C., SCHMOCKER U., DANG V.N. (eds.), *Probabilistic Safety Assessment and Management*, Vol. 4, Springer, London, 2004.

The power of scientists in up-stream EU policy-making

The scientific support, a necessary condition for a successful Commission initiative

From the EU energy economic research (See Chapter II), the energy-related policy case studies (See Chapter VI) and the specificity of the EU decision-making process (See Chapter VIII), it has been demonstrated that there are three necessary – but not sufficient – criteria in order to have a successful European Commission initiative, namely[68]:

- Economic and social agreement – or at least a wide consultation – of private, public and civil society organisations, i.e. the "end-users", on the content, timing and constraint of the policy initiative. The stakeholders should in a certain sense legitimate from the economic and social point of view the Commission proposal.
- Political consensus on common needs among the Commission (different DGs), the EU Council (commonalities amongst Member States) and the European Parliament (majority of votes often from several parties) that can be called the "genuine policy-makers";
- Scientific arguments on the future expected costs and benefits of the EU initiative provided by the "up-stream technocracy advisers": quantification and sometimes monetarisation of the impacts of the proposed objectives like the level of energy taxation, the targets for greenhouse gas emission reductions, for renewables, for energy efficiency or for the valuation of external costs.

The scientists should respect two values: the truth, i.e. that data are true and arrived at by scientific methods ("methodology filter"); and the authenticity, i.e. the "messenger" has integrity and makes claims in areas where he/she can be trusted (scientifically recognised institutions).

The scientific support to policies has been a precursor of the Impact Assessment that, since 2003, has to accompany all the major policy initiatives launched by the European Commission[69]. It tackles the issues of better regulation (direct costs and benefits of policy) and sustainable development (take account of wider effects – economic, social and environmental – of policy). It helps to identify the major problems to be solved, to outline the policy objectives; to consider the range of options and instruments.

---

[68] Successful means that the proposal will firstly be adopted by the Commission and later by the European Parliament and by the Council (co-decision) and finally become a European Directive.

[69] European Commission, *Communication on Impact Assessment*, COM(2002)276.

In a certain sense, scientific support to policies and Impact Assessment responds to the demands of subsidiarity and proportionality. It stimulates a consultation with stakeholders and with other policy DG. Finally, it defines the basis for the future implementation, monitoring and evaluation of the policy. The scientific support to policy and the Impact assessment is considered as a way of improving European governance (all policy-decisions should be based on sound analysis supported by the best data available)[70].

The credibility of the Commission energy-related policy initiatives is partly in the hand of modellers

The simulation results obtained with models served as a key background information and reference for the EU policy initiatives launched by the Commission these last fifteen years. Most of the preparatory and non-legally-binding European legislation was "scientifically supported" (see Chapter VI).

In the EU legislative process – generally a Commission Communication, followed by a (Commission) proposal for a Directive and finally by a European Parliament and Council Directive – the references and the evidences to the scientific support are only available in the unclassifiable legal preparatory documents that are the Commission Communications (including the Green and White Papers or the Proposal for a Directive looking in particular at the "Explanatory memorandums" and the "Annexes").

The accompanying information such as the Commission Staff Working Document or, since 2003 the Impact Assessment, are the most scientifically-based policy documents. These are not published in the Official Journal but by the Secretariat General of the Commission[71]. They set out a fuller technical and economic justification underlying the Commission position or the EU negotiation position and explain in details e.g. where the necessary reductions in greenhouse gas emissions are expected to take place, the likely economic impact of the measures and the possible instruments to implement the policies.

Model results were directly quoted in the Commission Green Paper on EU Emission Trading Scheme, in the Commission proposal for a Directive on the promotion of electricity produced from renewable

70 European Commission, Communication, *European Governance – a White paper*, COM(2001)428.

71 Before the development of information and communication technologies, these documents – not translated in all the EU languages – were not easily accessible to the public. These last years, with Internet and the new European Union "soft policy" where stakeholders are more informed during the whole decision-making process, the Staff working documents and Impact Assessments are rapidly available.

energy, and in the Commission Communication on energy efficiency in the EU. Models were also the main scientific sources in the Commission Staff Working Documents like those related to the Community framework for the taxation of energy products or for the Climate Change policy. More recently, models were the main source of the Impact Assessment concerning the Commission Communications related to Clean Air for Europe (socio-environmental damages or external costs) or to 2020 targets for renewable, energy efficiency and greenhouse gas emission reduction.

This is a clear indication of the policy value of the analysis performed by modellers, in particular GEM-E3, POLES, PRIMES, SAFIRE, MURE and EXTERNE (see Chapter V), all tools developed by EU energy economic research and that were largely quoted in the above-mentioned policies.

Already before the formal Impact Assessment (see Chapter VIII), EU energy-related policies were scientifically supported by modelling results. These "semi-formal evaluations" allowed an assessment of the expected economic, social, environmental and technological impacts of the proposed energy-related policy. With the Impact Assessment, the scientists (modellers) enter formally in the EU decision-making process.

The scientists are present in the whole EU energy-related decision-making process

The involvement of researchers is present at all stages of the EU policy-making cycle (*ex ante, in itinere* and *ex post*):

- Embracing the initial identification of the need for a policy action or response including foresight exercises, through academic and research projects (cf. Energy future outlooks);
- Shaping policy options, including through Impact Assessment (cf. use of various economic and/or energy modelling tools);
- Drafting policy proposals trough specific Commission tendered studies (cf. Communications, Green or White Papers, or Proposal for a Directive);
- Following the policy implementation (cf. Comitology including Member States representatives and experts);
- Monitoring and reviewing the policy (cf. *ex-post* assessment).

Following the EU energy-related policy case studies, one can conclude that the power of scientists in the EU decision-making process is:

- More important in the new policies like in the definition of the Emission Trading Scheme (cf. Green Paper of 2000 and the use of

POLES, PRIMES and GEM-E3) rather than in the traditional policies and measures.

- Higher in the policies where the data for the whole EU are not easily accessible and comparable like in the renewable or energy efficiency fields (cf. SAFIRE that was quasi exclusively used for drafting the Annex of the 2000 green electricity Directive proposal; and MURE that was quasi exclusively used for determining the energy efficiency potential in Europe).
- Only verifiable in the up-stream decision-making, i.e. at the level of the Commission Communications or Staff Working Documents (it cannot be detected in any of the EU legally-binding instruments, i.e. in any Regulation, Directive or Decision).
- More prevalent where there are important scientific aspects, e.g. resources and climate change.

META CONCLUSION

Science, Modelling and EU Policies

The seven scientists' challenges

What could be the role of scientists and more specifically of economic and energy modellers in supporting EU policies in the future?

Firstly, researchers should elaborate new and improved tools that are able to catch the increasing complexity of the world (cf. human behaviour patterns, "knowledge stock", intangible investments, innovation in services, geopolitical risks,), the very-long term problems (cf. resource depletion, climate change, radioactive waste management) and the specific links between economy, energy, and environment (cf. security of supply, externalities of energy, ecosystem damages).

Secondly, the methodologies and tools (like the Top-down and the Bottom-up models) should be compared and peer-reviewed, updated and completed. Economics, Energy and Environment (E3) models for example should be extended to include:

- the adaptation costs of climate change;
- transaction costs;
- non-CO_2 greenhouse gas emissions from *inter alia* agriculture, land-use and forestry;
- they should include actor analysis;
- they should address more accurately the question of new energy-efficient technologies at the end-use level;
- they should well cover multi-functionality (like distances between the production and consumption of goods, regional planning and infrastructures);
- they should try to be better integrated among themselves (economic and technological components – hybrid model) and, possibly, with large physical models (e.g. dealing with climate change).

Thirdly, modellers (both economists and engineers) should assist the whole decision-making process with appropriate tools. Researchers should have a continuous dialogue rather than an *ad hoc* relation with stakeholders and policy-makers. So involved in a specific stake and so engaged in policies, the most difficult for the scientist is to remain objective and rigorous, i.e. not being a "stake-holder" himself or not becoming the "policy-maker" himself. If that would be the case, it raises into question not only the credibility of the figures but the whole modelling community.

Fourthly, researchers should be present at all the assessment stages of a decision. In the *ex-ante* evaluation, to measure *a priori* the expected impacts from the policy. This independent and scientific position has a crucial role in the adoption of an EU initiative and in the trilogue among the Commission, the Council and the Parliament (cf. also the "Impact Assessment" that is required for all major EU political and programmatic initiatives since 2003).

The scientists should also be active in the *ex-post* evaluation that measures the impacts of the policy *a posteriori*. This helps policy-makers to decide if a policy should be renewed or extended. The evaluation *in itinere*, during the implementation of the policy, is probably the least known but certainly the one in which the results of the researchers could have the biggest impact.

Fifthly, scientists should work together. Economists and engineers but also physicists, political scientists, sociologists and psychologists should find ways to communicate better and collaborate. Multi-disciplinarity is not only an asset; it is a necessity. In fact, before being taken, decisions need to cross the different strata and concerns of the society. This is a key issue for policy-makers.

Sixthly, researchers should be at the forefront. They should present their results in the right time. They have to anticipate policy needs and not just follow decisions. If enlargement of Europe is a continuous dynamic since the 1970s, not many models appropriately cover the new Member States in due time. Hydrogen is a topic discussed for tens of years, but models start only recently – following the EU (and US) political impetus towards a hydrogen economy – to establish an "hydrogen module". In the same way, the formalisation of CO_2 capture and storage technologies within the energy models has often been requested by policy-makers rather than being anticipated by modellers.

Finally, researchers should reach a consensus – although this could challenge the notion of the "remarkable man". The divergence in modelling results is confusing to policy makers and, therefore, decreases the practical value of such models in real-world policy analysis. Policy-makers know that there is "divergence of opinions" or "uncertainty" concerning global warming, oil resources or renewables potential. But policy-makers can take into account scientific background only if it reflects a scientific consensus. In this respect, researchers should "take their responsibilities": to accept the results (even if they do not 100% agree) or to refuse them. The most difficult task for a policy-maker is to decide in a grey zone. Especially in the energy-related fields, nothing is completely black or white. But the policy-maker should be able to see clearly from the scientists if it is closer to black or closer to white. The example of climate change and the influence of scientists, especially through the IPCC is particularly illustrative in this regard.

Scientific consistency and research modesty

Scientific, quantitative and systemic tools (like E3 models) allow policy-makers to get a certain consistency. Prisoners of several challenges in the same time like security, competitiveness and sustainability in the energy field, policy-makers are in a certain sense "reassured" by science and by system analysis.

With quantitative and mathematical tools, policy-makers feel more independent from the diversified stakeholders direct or vested interests than with qualitative methods more connected to some personal, disciplinary or sectoral views, for example according to the choice of the experts interviewed in a Delphi survey.

Moreover, models allow accessing to the major assumptions that can be easily put into discussion (economic growth rate, oil price, demographic trends, etc.). Results for example, on the cost of nuclear or renewable technologies in 2030 can be easily identified. Hypotheses on the participation or not participation of developing countries to a post-Kyoto agreement can be immediately seen. The assumptions on industrialized countries efforts to reduce greenhouse gas emissions can be checked on a first glance.

Models only rarely provide quantitative convergent figures on the evaluations of policies. But, as mentioned by Paul Zagamé, they are very useful if the whole results ("concert des résultats") allow convincing the policy-makers on the qualitative nature of the phenomena.

Nevertheless, in a more and more globalized world, with a reinforcement of competition (merging and acquisition, productivity gains, use of information and communication technologies) and increasing flexibility (of work, of timing, of movement), modesty should be the key word about the relevance of the model results. Experience taught the need for a good deal of humility in analysing the future.

Whatever analytic tool or model is employed, the complexity of the future ensures that the consequences of certain policies can be predicted only with large uncertainty. Basic assumptions such as the future oil price could be wrong. Risks like climate change, or economic variables like the pace of decoupling between GDP and energy demand (the autonomous energy efficiency improvement) are easily under or overestimated. The social and political reactions are almost never anticipated.

Several foresight assessments have proved this statement: the Thomas Malthus' 18th century prediction that population growth would outrun the food supply; the 1970s Club of Rome warning of the imminent scarcity of a range of natural resources; or the "ENERGY 2000" study estimating a very small share of natural gas in the European Community electricity power generation.

Models allow for forecasting the whole energy system (supply and demand sides) in its economic context. They provide quantitative information for authorities dealing with long-term energy planning. Modelling studies nourishes policy analysis and provide users with energy scenarios that represent forecasts able to assess a specific policy case in comparison to a reference case. For example, based on results from models, the March 2007 European Council has fixed important targets for greenhouse gas emissions (20% reduction by 2020 in comparison to 1990), for energy efficiency (20% energy savings by 2020 in comparison to the reference case) and for renewable energy sources (20% in the energy mix by 2020 compared to almost 7% in 2006).

In one word, the scientific consistency is necessary to policy-makers but the research results should always be regarded with modesty.

Democracy and technocracy in EU policy-making

The following debate is at stake: democracy and transparency on one side *versus* technocracy and "operativity" on the other side.

The European Community – as many other public institutions – will continue to struggle with a critical dilemma: is it better to have a common and transparent modelling platform (open, user-friendly, etc.) but remaining very general in its answers? Or should it continue to have a set of detailed and rich models – rather complex and opaque – able to answer to technical and specific political questions?

Looking from a macro-institutional perspective, the European Parliament – representing the citizens – is more favourable to the first option while the European Commission and its "technical charisma" – prefers the second one. The debate is between democracy and technocracy, people and stakeholders, general value and specific problems.

Without being democratic like the Parliament (elected since 1979) and without assuming the national legitimacy of the Council, the Commission uses science, facts and figures, to base its authority, in particular in its right of initiative, i.e. when making policy proposals. For example, the economic assumptions made by the experts in the various Top-down and Bottom-up models used in EU energy-related policies play a key role in the final outputs of the models but are only marginally questioned in the purely "political debate" (Parliament and Council) while they are strongly discussed between the experts and the Commission in the "techno-political" debate.

The role of the Commission is certainly higher in more technical and economic matters – like trade policy or energy-related policies – than on social and political matters (employment, education, defence, etc.). In the techno-economic matters, the Commission has a key role in interpreting and translating the scientific issues (like the model functioning and results) into political reality. As underlined by Catherine Day, the Commission Secretary General following the European Council agreement on the climate change and energy package (cf. 2020 targets) of 11-12 December 2008, "The very careful preparation of these proposals has paid off: the professionalism and technical proficiency of the Commission has proved capable of resisting the inevitable pressure to water down the proposals in the face of economic crisis".

The role of scientists in EU policy-making is being more and more formalised. From the top of the European legal acts, i.e. the EU Treaty (already since 1986 for environment policy in the Single European Act) to the principles and guidelines on the use of expertise in the EU passing by the good Governance principles, the expert's committees that accompany the Commission in the execution of delegated powers ("Comitology"), and the procedure of Impact Assessment, the scientists participate more and more in the EU decision-making process.

The relation between experts and the Commission can be characterised as a technical rationality model that gives priority to the links between the experts and the deciders. Nevertheless, as underlined by Philippe Roqueplo, these two ways of deciding are not sufficient and an equilibrium should be found among the three sides of the triangle, and therefore also between the experts and the citizens.

In a few words, this book has highlighted the role of the "experts" in the EU decision-making process. It has especially focused on the "origin of the figures" (target, cost, potential) in the energy-related policies. Scientists – mainly economists and modellers – do not limit their role to make recommendations to the Commission. They are fully involved in the EU policy-making process from the early beginning of the decision process. Scientists are really co-shapers of European decisions. This is maybe not amenable for all the policy areas, but it is a fact for the EU energy and environment policies.

Glossary and Acronyms

AEEI	Autonomous Energy Efficiency Improvement
Bl	Barrel
BRUEGEL	Brussels European and Global Economic Laboratory
CAP	Common Agricultural Policy
CEPR	Centre for Economic Policy Research
CEPS	Centre for European Policy Studies
CHP	Cogeneration of Heat and Power
CGE	Computable General Equilibrium Model
DICE	Dynamic Integrated model of Climate and the Economy
EASW	European Awareness Scenario Workshop
E3ME	Energy-environmental-economy model for Europe
E3MG	Energy, Environment, Economy Model of the Globe
EC	European Community(ies)
ECIPE	European Centre for International Political Economy
ECTS	European Credit Transfer System
EEIG	European Economic Interest Grouping
EFL	Electricity Feed-in Law
EFOM	Energy Flow Optimisation Model
EMF	Energy Modelling Forum
EP	European Parliament
EPC	European Policy Centre
ERA	European Research Area
EREC	European Renewable Energy Council
ETC	Endogenous Technical Change
ETS	Emission Trading Scheme
ETSAP	Energy Technology Systems Analysis Programme
EU	European Union
FP	Framework Programme
GEM-E3	General Equilibrium Model for Energy, Economy and Environment

GHG	Greenhouse gas emissions
GJ	Gigajoule
GTAP	Global Trade Analysis Project
GW	Gigawatt
GWh	Gigawatthour
HERMES	Harmonised Econometric Research for Modelling Economic Systems
IA	Impact Assessment
IEA	International Energy Agency
IEO	International Energy Outlook
IIASA	International Institute for Applied for Systems Analysis
IMAGE	Integrated Model to Assess the Global Environment
IMCP	Innovation Modelling Comparison Project
I/O	Input/Output
IPCC	Intergovernmental Panel on Climate Change
IPR	Intellectual Property Right
LP	Linear Programming
MARKAL	Market Allocation
MEDEE	Modèle de prospective de la Demande Énergétique à long terme
MESSAGE	Model for Energy Supply Strategy Alternatives and their General Environmental impact
MIDAS	Managing Input Data System
MJ	Megajoule
Mt CO_2	Million tonnes of carbon dioxide
Mtoe	Million tonnes of oil equivalent
MURE	Maîtrise et Utilisation Rationnelle de l'Énergie
MW	Megawatt
NEMS	National Energy Modelling System
OJ	Official Journal
POLES	Prospective Outlook for the Long-term Energy System
PRIMES	Price Induced Model of the Energy System
R&D	Research and Development
Toe	Tonne of oil equivalent
SAGE	System for the Analysis of Global Energy Markets

SAFIRE	Strategic Assessment Framework for the Implementation of Rational Energy
TSER	Targeted Socio-Economic Research
UNFCCC	United Nations Framework Convention on Climate Change
WEC	World Energy Council
WEM	World Energy Model
WEO	World Energy Outlook
WETO	World Energy Technology Outlook

List of Figures

List of Tables

Bibliography

ACHACHE J., *Les sentinelles de la terre*, Hachette, Paris, 2004.

AGHION P, HOWITT P., *Endogenous growth theory*, MIT Press, Cambridge MA, 1998.

ALBERTH S., HOPE C., *Climate modelling with endogenous technical change: stochastic learning and optimal greenhouse gas abatement in the PAGE2002 model*, Energy Policy, Vol. 35, March 2007.

ALLEMAND F., *L'Union pour la Méditerranée: Pourquoi? Comment?*, Fondation pour l'innovation politique, Paris, juin 2008.

ARROW K., *The economic implications of learning by doing*, The Review of Economic Studies, Vol. 29, No. 3, June 1962.

ARTIS M., NIXSON F., *The economics of the European Union*, Oxford University Press, 2007.

AUDIGER P., *La politique énergétique de la Commission européenne et la concurrence: un éclairage politique*, Revue de l'Énergie, No. 577, Paris, mai-juin 2007.

BATTLE A., *Les travailleurs du futur*, Seghers, Paris, 1986.

BERGER G., *L'attitude prospective*, Revue prospective, No. 1, 1958.

BIJKER W., HUGHES T., PINCH T., *The social construction of technology systems*, MIT Press, Cambridge, 1987.

BIJKER W., LAW J., *Shaping technology/building society: studies in sociotechnical change*, The MIT Press, Cambridge Ma, 1994.

BJERREGAARD R., *Speech on environmental tax reform*, Conference on Environmental tax reform in Europe, Brussels, 27/10/1997.

BLONDEL D., *L'innovation pour le meilleur et pour le pire*, Hatier, Paris, 1990.

BOEHRINGER C. *et al.*, *European emission mitigation policy and technological evolution: economic evaluation with the GEM-E3*, 2000, Final Report to the European Commission (DG Research).

BOEHRINGER C., LOSCHEL A., *Assessing the costs of compliance: the Kyoto Protocol*, European Environment, Wiley, Vol. 12, January-February. 2002.

BOEHRINGER C., *The synthesis of Bottom-up and Top-down in energy policy modelling*, Energy Economics, No. 20, 1998.

BOITEUX M., BAUMSTARK L., *Transports: choix des investissements et coûts des nuisances*, La Documentation française, Paris, 2001.

BOURGEOIS B., FINON D., MARTIN J.M., *Énergie et changement technologique – une approche évolutionniste*, Economica, Paris, 2000.

BOURINET J., SNYDER F., *La sécurité alimentaire dans l'Union européenne*, Bruylant, Bruxelles, 2003.

BRANSCOMB L., HOTON G., SONNERT G., *Science for Society, Basic research in the service of public objectives*, Packard and Sloan Foundations, 2001.

BRUCE J.P., LEE H., HAITES E.F., *Climate Change: Economic and social dimensions of climate change*; Contribution of Working Group III to the Second Assessment Report of the Intergovernmental Panel on Climate Change, Cambridge University Press, UK, 1995.

Bundesministerium für Bildung und Forschung, *Futur: Der deutsche Forschungsdialog*, BMBF, Bonn, 2003.

Bureau Fédéral du Plan, *Perspectives énergétiques pour la Belgique à l'horizon 2030*, Bruxelles, 2005.

CAMOS M., DUMOAT A., VALETTE P., *MEDEE 3 – Modèle de demande en énergie pour l'Europe*, EC-TEC-DOC, Paris-Luxembourg, 1986.

CAPROS P., MANTZOS L., *The economic effects of EU-wide industry-level emission trading to reduce greenhouse gases*, European Commission, DG Research, Conference on climate negotiations and emission trading, Brussels, 29-30/08/2000.

CAPROS P., *Energy technology dynamics and advanced energy systems modelling*, International Journal of Global Energy Issues, United Kingdom, 2000.

CAPROS P., GEORGAKOPOULOS T., MANTZOS L., *Economic and energy system implications of European CO_2 mitigation strategy for 2010: a model based analysis*, National Technical University of Athens, 1997.

CAPROS P., GEORGAKOPOULOS T., VAN REGEMORTER D., PROOST S., SCHMIDT T., KOSCHEL H., CONRAD K., VOUYOUKAS E.L, *Climate Technology Strategy 2*, Physica-Verlag, Germany, 1999.

CAPROS P., KOUVARITAKIS N., MANTZOS L., PANOS V., VOUYOUKAS L., *Note on the impact of fuel taxation on technology choice*, November 2000, Study for the Commission (DG TREN).

CAPROS P., MANTZOS L., CRIQUI P., KOUVARITAKIS N., SORIA RAMIREZ A., SCHRATTENHOLZER L., VOUYOUKAS E.L., *Climate Technology Strategy 1*, Physica-Verlag, Germany, 1999.

CARTOU L., CLERGEVIE J.-L., GRUBER A., RAMBAUD P., *L'Union européenne*, Dalloz, Paris, 2006.

Center for International Climate and Environmental Research (CICERO), *Climate change negotiations: Burden sharing and cost-effective implementation mechanisms and protocols – A conference summary*, TemaNord, 562, Denmark, 1997.

CHALTIEL F., *Le processus de décision dans l'Union européenne*, La Documentation française, Paris, 2006.

CHATEAU B., HAMACHER T., SMEKENS K., MARTINSEN D., REISINGER H., PATEL M., *Very long term energy modelling (VLEEM)*, Final report, European Commission (DG RTD), Brussels, 2000.

CHATEAU B., LAPILLONNE B., *Long-term energy demand simulation*, in A. (ed.), *Energy models for the European Community*, IPC Science and Technology Press Limited, 1979.

CHATEAU B., LAPILLONNE B., *Energy demand: facts and trends – a comparative analysis of industrialized countries*, Springer-Verlag, Vienna, 1982.

CHEVALIER J.M., *Les grandes batailles de l'énergie*, Gallimard, Paris, 2004.

CHEVALLIER C., FOUGEYROLLAS A., LE MOUEL P., ZAGAME P., *A time to sow, a time to reap for the European countries: a macro-econometric glance at the RTD national action plans*, Revue OFCE, June 2006.

CHERMACK T.J., LYNHAM S.A., RUANA W.E.A., *A review of scenario planning literature*, Futures Research Quarterly, Vol. 17, No. 2, 2001.

CHRISTIANSEN T., LARSSON T., *The role of committees in the European-Union policy-process*, Edward Elgar, 2007.

CHRISTIANSEN T., VACCARI B., *The 2006 Reform of Comitology: Problem solved or dispute postponed*, EIPASCOPE No. 3, Maastricht, 2006.

CNRS/LEPII, *Greenhouse gas reduction pathways in the UNFCCC process up to 2025*, Study for the Commission (DG Environment), October 2003.

COHERENCE, *Economic evaluation of quantitative objectives for Climate Change*, Study for the Commission (DG Environment), 1999.

COLLINS P., LINDBERG U., *Independent science advice in the European Union*, IPTS Report, No. 70, December 2002

Commissariat Général du Plan, *Énergie 2010-2020*, La Documentation française, Paris, 1998.

Commission européenne, *Europe de l'Énergie en 2020 – Une approche par scénario*, OPOCE, Luxembourg, 1996.

CORBETT R., JACOBS F., SHACKLETON M., *The European Parliament*, John Harper, UK, 2007.

Council *Decision adopting a specific programme for research, technological development and demonstration: "Integrating and strengthening the ERA (2002-2006)*, OJ, L 294, 29/10/2002.

Council *Decision concerning the framework programme for Community RTD activities (1987-1991)*, OJ, 24/10/1987.

Council *Decision concerning the framework programme for Community RTD activities (1990-1994)*, OJ, 8/5/1990.

Council *Decisions 87/373/EEC, 1999/468/EC, 2006/512 laying down procedures for the exercise of implementing powers conferred on the Commission*, OJ L 197, 18/7/1987, OJ L 184, 17/7/1999, OJ L 200, 22/7/2006.

Council *Decision on a European strategic programme for research and development in information technologies (ESPRIT)*, OJ, L 67, 9/3/1984.

Council *Decision on a plan to stimulate European cooperation and exchange in the scientific and technical field*, OJ, L 83, 25/3/1985.

Council *Decision on a programme of research activities to be executed by the Joint Research Centre for the EAEC and EEC (1984-1987)*, OJ, L 3, 5/1/1984.

Council *Directive 2003/96/EC restructuring the Community framework for the taxation of energy products and electricity*, OJ, L 283, 27/10/2003.

Council *Directive 92/81/EEC on the harmonization of the structures of excise duties on mineral oils*, OJ, L 316, 31/10/1992.

Council *Directive 92/82/EEC on the approximation of the rates of excise duties on mineral oils*, OJ, L 316, 31/10/1992.

Council *Directive 99/30/EC relating to limit values for sulphur dioxide, nitrogen dioxide and oxides of nitrogen, particulate matter and lead in ambient air*, OJ, L 163, 29/6/1999.

European Council, *Presidency conclusions*, Brussels, 8-9 March 2007.

Council of the European Union, *Review of the EU sustainable development strategy – Renewed strategy*, 9/6/2006.

Council *Resolution concerning a first action programme of the European Communities in the field of science and technology*, OJ, C 7, 29/1/1974.

Council *Resolution concerning an action programme of the European Communities in the field of forecasting, evaluation and methodology*, OJ, C 7, 29/1/1974.

Council *Resolution concerning FP for Community RTD&D and FP1 (1984-1987)*; OJ, 4/8/1983.

Council *Resolution concerning the coordination of national policies and the definition of projects of interest to the Community in the field of science and technology*, OJ, C 7, 29/1/1974.

Council *Resolution concerning the participation of the European Communities to the European Science Foundation*, OJ, C 7, 29/1/1974.

Council *Resolution on a programme of support for technological development in the hydrocarbons sector*, OJ, L 350, 27/12/1985.

Council *Resolution on an initial outline programme of the EC in the field of science and technology*, OJ, C 7, 29/1/1974.

Council *Resolution on the greenhouse effect and the Community*, OJ, C 183, 21/6/1989.

Council *Resolution on the promotion, by financial support, of demonstration projects and industrial projects in the energy field*, OJ, L 350, 27/12/1985.

COURTNEY H., *20/20 Foresight: Crafting strategy in an uncertain world*, Harvard Business School Press, Massachusetts, 2002.

CRIQUI P., MIMA S. and VIGUIER L., *Marginal abatement costs of CO_2 emission reductions, geographical flexibility and concrete ceilings: an assessment using the POLES model*, Energy Policy, 27, 1999.

CRIQUI P., VIGUIER L., *The costs of CO_2 reduction and the impact of flexibility mechanisms in meeting the Kyoto targets*, Int. J. Global Energy Issues, 2000.

DAHAN DALMEDICO A. (ed.), *Les modèles du futur – Changement climatique et scénarios économiques: enjeux scientifiques et politiques*, Recherches, La Découverte, Paris, 2007.

DAS A., ROSSSETTI di VALDALBERO D., VIRDIS M., *ACROPOLIS: An example of international collaboration in the field of energy modelling to support greenhouse gases mitigation policies*, Energy Policy, Vol. 35, No. 2, February 2007.

DAY C., *Note from the Secretary General on the 11-12 December 2008 European Council*, Brussels, 12/12/2008.

DECOUFLE A.C., *Traité élémentaire de prévision et de prospective*, PUF, Paris, 1978.

DEFRAIGNE J.-C., *Économie européenne et politique*, L'Harmattan, Paris, 2004.

DE LAAT B., *Sociology for energy modelling and the sociology of energy models*, in GIOVANNINI B., BARANZANI A., *Energy modelling beyond economics and technology*, CUEPE, Genève, 1997.

DE MAN R., *Energy forecasting and the organization of the policy process*, Eburon, Delft, 1987.

DENZIN N.K., LINCOLN Y.S., *The landscape of qualitative research*, Sage publications, California, 1998.

DE JOUVENEL H., *Invitation à la prospective*, Perspectives, Futuribles, 2004.

DESSELAS S., *Un lobbying professionnel à visage découvert*, éd. du Palio, 2007.

DETEMMERMAN V., DONNI E., ZAGAME P., *Increase of taxes on energy as a way to reduce CO_2 emissions: problems and accompanying measures*, Report for the European Communities, DG XII, July 1991

DOWLATABI H., *Integrated assessment of climate change: An incomplete overview*, in NAKICENOVIC N., NORDHAUS W.D., RICHELS R., TOTH F.L., *Integrative assessment of mitigation, impacts, and adaptation to climate change*, ed. International Institute for Applied Systems Analysis (IIASA), Laxenburg-Austria, 1994.

DUPUY J.P., *Pour un catastrophisme éclairé*, Seuil, Paris, 2002.

DURAND J., *Prospective, discontinuité et instabilité*, Futuribles, Paris, Automne 1975.

DURANTE B., *Analisi dei costi esterni non corrisposti da chi effettua il trasporto nel corridoio Bolzano-Brennero*, Bolzano, 2001.

DYSON K., FEATHERSTONE K., *The road to Maastricht: Negotiating Economic and Monetary Union*, Oxford University Press, 1999.

ECMWF, *World Modelling Summit for Climate Prediction*, Reading (UK), 6-9 May, 2008.

ECOFYS, *Economic evaluation of sectoral emission reduction objectives for Climate Change*, Study for the Commission (DG Environment), March 2001.

EDMONDS, J.A., WISE M.A., MacCRACKEN C.N., *Advanced energy technologies and climate change: An analysis using the global change assessment model (GCAM)*, Pacific Northwest Laboratory, Richland, Washington, 1994.

EIA, *Annual energy outlook – Retrospective review: Evaluation of projections in past editions (1982-2006)*, Washington, 2006.

ELLERMAN D., SCHMALENSEE R., JOSKOW P., MONTERO J.-P., BAILEY E., *Markets for clean air: the US acid rain program*, Cambridge University Press, Cambridge and New-York, 2000

ELLUL J., *La technique ou l'enjeu du siècle*, Armand Colin, Paris, 1954.

ELLUL J., *Le système technicien*, Calman-Lévy, Paris, 1977.

ELSASSER W., *Atom and organism: A new approach to theoretical biology*, Princeton University Press, 1966.

Energy and Environmental Council, *EC climate change policy*, Luxembourg, 29/10/1990.

Energy Research Centre of the Netherlands, *The next 50 years: Four European energy futures*, Petten, 2005.

EREC/GREENPEACE, *Energy (r)evolution – a sustainable world energy outlook*, EREC, The Netherlands, 2007.

EURELECTRIC, *Pricing the environment? An update on ExternE, Proceedings*, Brussels, 2001.

European Commission, *Action Plan for energy efficiency: Realising the potential*, COM(2006)545 and Staff Working Document, SEC (2006)1173.

European Commission, *Action plan on simplifying and improving the regulatory environment*, COM(2002)278.

European Commission, *Action plan to improve energy efficiency in the European Community*, COM(2000)247.

European Commission, *Amended proposal for a Directive introducing a tax on carbon dioxide emissions and energy*, COM(95)172.

European Commission, *An Energy Policy for Europe*, COM(2007)1.

European Commission, *A strategic review of better regulation in the European Union*, COM(2006)689.

European Commission, *A strategy for integrating the environment into EU policies*, COM(98)333.

European Commission, *A sustainable Europe for a better world: A European strategy for sustainable development*, COM(2001)264.

European Commission, *A European strategic energy technology plan (SET-Plan) – Towards a low carbon future*, COM(2007)723.

European Commission, *Clean Air for Europe (CAFE) Programme*, COM(2001)245.

European Commission, Staff working document, *EU energy policy data*, SEC(2007)12.

European Commission, *Communication on an overall view of energy policy and actions*, COM(97)167.

European Commission, *Communication on climate change – The EU approach for Kyoto*, COM(97)481.

European Commission, *Communication on climate change – Towards a EU post-Kyoto strategy*, COM(98)353.

European Commission, *Communication on cohesion and research policy: synergies between research and technology development policy and cohesion policy*, COM(93)203.

European Commission, *Communication on Community policy targets on greenhouse issue*, SEC(90)496.

European Commission, *Communication on Impact Assessment*, COM(2002)276.

European Commission, *Communication on the collection and use of expertise by the Commission: principles and guidelines – Improving the knowledge base for better policies*, COM(2002)713.

European Commission, *Communication on the regional dimension of the European research Area*, COM(2001)549.

European Commission, *Communication on transport and CO_2 – Developing a Community approach*, COM(98)204.

European Commission, *Community guidelines on state aid for environmental protection*, OJ, C 37, 3/2/2001.

European Commission, *Community strategy to combat acidification*, COM(97)88.

European Commission, *Community strategy to limit carbon dioxide emissions and improve energy efficiency*, COM(92)246 and SEC(91)1744.

European Commission, DG Research, *Key figures 2005*, Brussels, 2006

European Commission, DG XII, *Cost-effectiveness analysis of CO_2 reduction options*, Brussels, 1991.

European Commission, DG XII, *Increase of taxes of energy as a way to reduce CO_2 emissions: problems and accompanying measures*, Brussels, 1991

European Commission, *Directions for the EU on environmental indicators and green national accounting – The integration of environmental and economic information systems*, COM(94)670.

European Commission, *Energy – Economic, social and policy-oriented research in Europe*, OPEC, Luxembourg, 2007.

European Commission, *Energy and transport trends to 2030*, OPEC, Luxembourg, 2003.

European Commission, *Energy and transport: scenarios on energy efficiency and renewables*, OPEC, Luxembourg, 2006.

European Commission, *Energy cooperation with the developing countries*, COM(2002)408.

European Commission, *Energy dimension of climate change*, COM(97)196.

European Commission, *Energy efficiency in the European Community – towards a strategy for the rational use of energy*, COM(98)246.

European Commission, *Energy futures – The role of research and technological development*, Luxembourg, 2006.

European Commission, *EU policies and measures to reduce greenhouse gas emissions: Towards a European climate change programme (ECCP)*, COM(2000)88.

European Commission, *European climate change programme (ECCP)*, Report, Brussels, June 2001.

European Commission, *European governance – a White paper*, COM(2001)428.

European Commission, *European technology platforms – Knowledge for growth*, EC, Brussels, 2005.

European Commission, *European transport policy for 2010: time to decide*, COM(2001)370.

European Commission, *European Union Energy Outlook to 2020*, EC, Belgium, 1999.

European Commission, *ExternE: Externalities of Energy, Summary* (Vol. 1), *Methodology* (Vol. 2), *Coal and lignite* (Vol. 3), *Oil and gas*, (Vol. 4), *Nuclear* (Vol. 5), *Wind and hydro* (Vol. 6), OPEC, Luxembourg, 1995.

European Commission, *ExternE: Externalities of Energy, Methodology 1998 update*, (Vol. 7), *Global warming* (Vol. 8), *Fuel cycles for emerging and end-use technologies, transport and waste* (Vol. 9), *National Implementation* (Vol. 10), OPEC, Luxembourg, 1999.

European Commission, *ExternE: Externalities of Energy, Methodology 2005 update*, OPEC, Luxembourg, 2005.

European Commission, *Five year assessment report related to the specific programme "Energy, Environment and Sustainable Development"*, Brussels, June 2000.

European Commission, Forward Studies Unit, *Scenarios Europe 2010 – Five possible futures for Europe*, Brussels, 1999.

European Commission, *GEM-E3 computable general equilibrium model for studying economy-energy-environment interactions*, OPEC, Luxembourg, 1995.

European Commission, Green Paper, *Adapting to climate change in Europe – options for EU action*, COM(2007)354.

European Commission, *Green Paper on a European strategy for sustainable, competitive and secure energy*, COM(2006)105.

European Commission, *Green Paper on energy efficiency or doing more with less*, COM(2005)265.

European Commission, *Green Paper on greenhouse gas emissions trading within the European Union*, COM(2000)87.

European Commission, Green Paper, *Energy for the future: Renewable sources of energy*, COM(96)576.

European Commission, Green Paper, *The European Research Area: New perspectives*, COM(2007)161.

European Commission, *Hydrogen energy and fuel cells – A vision of our future*, EC, Brussels, 2003.

European Commission, *Implementing the Community strategy – Reduce CO_2 emissions from cars: An environmental agreement with the European automobile industry*, COM(98)495.

European Commission, *Integrating environmental considerations into other policy areas – a stocktaking of the Cardiff process*, COM(2004)394.

European Commission, *JOULE I – catalogue of contracts, programme 1989-1992*, OPEC, Luxembourg, 1993.

European Commission, *JOULE II – project synopses, programme 1990-1994*, OPEC, Luxembourg, 1994.

European Commission, *JOULE III – project synopses, programme 1994-1998*, OPEC, Luxembourg, 1999.

European Commission, *Limiting global climate change to 2 degrees Celsius – The way ahead for 2020 and beyond*, COM(2007)2.

European Commission, *Making a reality of the European Research Area: guidelines for EU research activities*, COM(2000)612.

European Commission, *More Research for Europe: Towards 3% GDP*, COM(2002)499.

European Commission, *Participating in European research*, Brussels, 2004.

European Commission, *POLES*, OPEC, Luxembourg, 1996.

European Commission, *Preparing for implementation of the Kyoto Protocol*, COM(99)230.

European Commission, *PRIMES*, OPEC, Luxembourg, 1995.

European Commission, *Proposal for a Directive amending Directive 88/609/EEC on the limitation of emissions of certain pollutants into the air from large combustion plants*, COM(1988)415.

European Commission, *Proposal for a Directive amending Directive 99/62/EC on the charging of heavy goods vehicles for the use of certain infrastructures*, COM(2003)448.

European Commission, *Proposal for a Directive establishing a scheme for greenhouse gas emission allowance trading within the Community and amending Council Directive 96/61/EC*, COM(2001)581.

European Commission, *Proposal for a Directive on energy end-use efficiency and energy services*, COM(2003)739.

European Commission, *Proposal for a Directive on establishing a framework for the setting of Eco-design requirements for Energy-Using Products and amending Council Directive 92/42/EC*, COM(2003)453.

European Commission, *Proposal for a Directive on the energy performance of buildings*, COM(2001)226.

European Commission, *Proposal for a Directive on the geological storage of carbon dioxide*, COM(2008)18.

European Commission, *Proposal for a Directive on the promotion of electricity produced from renewable energy sources in the internal electricity market*, COM(2000)279.

European Commission, *Proposal for a Directive restructuring the Community framework for the taxation of energy products*, COM(97)30.

European Commission, *Proposal for a Directive restructuring the Community framework for the taxation of energy products*, COM(97)30.

European Commission, *Renewable energy road map – Renewable energies in the 21st century: building a more sustainable future*, COM(2006)848.

European Commission, *Renewable energy road map – Renewable energies in the 21st century: building a more sustainable future*, Staff Working Document, SEC (2006)1719.

European Commission, *SAFIRE*, OPEC, Luxembourg, 1995.

European Commission, *Scenarios to year 2000: a prospective study of the energy market in the European Community*, Internal Working Paper, 1983.

European Commission, *Second European Report on S&T indicators 1997*, OPEC, Luxembourg, 1997.

European Commission, *Simplifying and improving the regulatory environment*, COM(2002)278.

European Commission, *Sixth Environment Action Programme – Environment 2010: our future, our choice*, COM(2001)31.

European Commission, *Socio-economic projects in energy and environment*, OPEC, Luxembourg, 2001.

European Commission, *Socio-economic tools for sustainability impact assessment*, OPEC, 2002.

European Commission, *Staff Working Document accompanying the Communication Limiting Global Climate Change to 2 degrees Celsius – The way ahead for 2020 and beyond*, SEC(2007)8.

European Commission, *Staff Working Document accompanying the Green Paper, The European Research Area: New perspectives*, SEC(2007)412.

European Commission, Staff Working Document, *Annex to the proposal for the Council and European Parliament decisions on the seventh EU RTD Framework Programmes (EC and Euratom)*, COM(2005)119 and SEC(2005)430.

European Commission, Staff Working Document, *The share of renewable energy in the EU*, SEC(2004)547.

European Commission, *Staff Working Paper on the Thematic strategy on air pollution*, SEC(2005)1132; and *Impact Assessment of the Thematic strategy and the Clean Air For Europe (CAFÉ) Directive*, SEC (2005)1133.

European Commission, Staff Working Paper, *Extended impact assessment on the Directive amending the Directive establishing a scheme for greenhouse gas emission allowance trading within the Community, in respect of the Kyoto Protocol's project based mechanisms*, SEC(2003)785.

European Commission, Staff Working Paper, *Winning the battle against global Climate Change*, SEC(2005)180.

European Commission, *Stimulating technologies for sustainable development: An Environmental Technologies Action Plan for the European Union*, COM(2004)38.

European Commission (DG TAXUD), *Study on economic and environmental impacts of energy taxation*, 2003.

European Commission study (DG TREN), *Handbook on estimation of external costs in the transport sector*, Delft, 2008.

European Commission, *Tax policy in the European Union – Priorities for the years ahead*, COM(2001)260.

European Commission, *The energy dimension of Climate Change*, COM(97)196.

European Commission, *The share of renewable energy in the EU*, COM(2004)366.

European Commission, *Thematic strategy on air pollution*, COM(2005)446.

European Commission, *Third European report on Science and Technology indicators*, OPEC, Luxembourg, 2003.

European Commission, *Towards a European Research Area*, COM(2000)6.

European Commission, *Towards a European Strategic Energy Technology Plan (SET-Plan)*, COM(2006)847.

European Commission, *Towards a European strategic programme for research and development in information technologies*, COM(82)287.

European Commission, *Towards a European strategy for the security of energy supply*, COM(2000)769.

European Commission, *Towards a reinforced culture of consultation and dialogue – General principles and minimum standards for consultation of interested parties by the Commission*, COM(2002)704.

European Commission, *Towards sustainability: a European Community programme of policy and action in relation to the environment and sustainable development*, COM(92)23.

European Commission, *White Paper on European transport policy for 2010*, COM(2001)370.

European Commission, *White Paper on growth, competitiveness and employment: the challenges and ways forward into the 21st century*, COM(93)700.

European Commission, *An energy policy for the European Union*, COM(95)682.

European Commission, *Energy for the future – Renewable sources of energy*, COM(97)599.

European Commission, *Winning the battle against global Climate Change*, COM(2005)35.

European Commission, *Work Programme for the specific programme for RTD&D "Integrating and strengthening the ERA", Specific activities covering a wider field of research, Policy support and anticipating scientific and technological needs*, 9/12/2003.

European Commission, Working Paper of the Commission departments, *Presentation of the new Community system for the taxation of energy products, Assessment of the impact of the proposal*, SEC (97)1026/2.

European Commission, *World energy, technology and climate policy outlook (WETO-2030)*, Luxembourg, 2003.

European Commission, *World Energy Technology Outlook-2050 (WETO-H2)*, Luxembourg, 2006.

European Environment Agency, *Environmental signals 2001*, Copenhagen, 2001.

European Parliament and Council *Decision concerning the fifth framework programme of the EC for RTD&D activities (1998-2002)*, OJ, 1/2/1999.

European Parliament and *Council Decision concerning the fourth framework programme of the EC for RTD&D activities (1994-1998)*, OJ, 18/5/1994.

European Parliament and Council *Decision concerning the sixth framework programme of the EC for RTD&D, contributing to the creation of the ERA and to innovation (2002-2006)*, OJ, 29/8/2002.

European Parliament and Council *Directive 2001/42/CE on the assessment of the effects of certain plans and programmes on the environment*, OJ L 197, 21/07/2001.

European Parliament and Council *Directive 2001/77/EC on the promotion of electricity produced from renewable energy sources in the internal electricity market*, OJ, L 283, 27/10/2001.

European Parliament and Council *Directive 2004/35/EC on environmental liability with regard to the prevention and remedying of environmental damage*, OJ, L 143, 21/04/2004.

European Parliament and Council *Directive 2006/38 amending Directive 1999/62/EC on the charging of heavy good vehicles for the use of certain infrastructures*, OJ, L 157, 9/6/2006.

European Parliament and Council *Directive 2002/91/EC on energy performance of buildings*, OJ, L1, 4/1/2003.

European Parliament and Council *Directive 2003/30/EC on the promotion of biofuels*, OJ, L 123, 17/5/2003.

European Parliament and Council *Directive 2003/87/EC establishing a scheme for greenhouse gas emission allowance trading within the Community and amending Council Directive 96/61/EC*, OJ, L 275, 25/10/2003.

European Parliament and Council *Directive 2004/8/EC on the promotion of cogeneration*, OJ, L 52, 21/2/2004.

European Parliament and Council *Directive 2005/32/EC establishing a framework for the setting of eco-design requirements for energy-using products and amending Directives 92/42/EEC*, 96/57/EC and 2000/55/EC, OJ, L191, 22/7/2005.

European Parliament and Council *Directive 2006/32/EC on energy end-use efficiency and energy services and repealing Directive 93/76/EEC*, OJ, L 114, 27/4/2006.

European Parliament and Council *Directive 2001/80/EC on the limitation of emissions of certain pollutants into the air from large combustion plants*, OJ, L 309, 27/11/2001.

European Parliament and Council *Directive 2001/81/EC on national emission ceilings for certain atmospheric pollutants*, OJ, L 309, 27/11/2001.

European Parliament, *Resolution on measures to combat rising CO_2 concentrations in the atmosphere*, OJ, C 255, 13/10/1986.

European Union, *Consolidated versions of the Treaty on European Union and of the Treaty establishing the European Community*, OJ, C 321, 29/12/2006.

Eurostat, *Integration – Indicators for energy*, OPEC, Luxembourg, 2000.

FERNANDEZ ESTEBAN M., *The rule of law in the European Constitution*, Kluwer, The Hague, 1999.

FINON D., *Prospective énergétique et modélisation – Identification de pistes de progression méthodologique*, Note au Conseil Scientifique de l'Institut Français de l'Énergie, Paris, juin 2003.

FINON D., *Prospective énergétique et modélisation de long terme: les voies de progression méthodologique*, Revue de l'Énergie, No. 553, janvier 2004.

FINON, D., *Scope and limitations of formalized optimization of a national energy system*, in A. STRUB (ed.), *Energy models for the European Community*, IPC Science and Technology Press Limited, 1979.

Foundation for International Environmental Law and Development in London, *Designing options for implementing an Emissions Trading regime, for greenhouse gases in the EC*, Study for the Commission (DG Environment), February 2000.

FISCHER BOEL M., *Food, feed or fuel: a measured policy on agricultural markets*, Meeting with Bundesvereinigung der Deutschen Ernährugsindustrie, Berlin, 18 January 2008.

FORRESTER J.W., *Principles of Systems*, Wright-Allen Press, Cambridge MA, 1968.

FORRESTER J.W., *World Dynamics*, Wright-Allen Press, Cambridge MA, 1971.

FOUGEYROLLAS A., LE MOUEL P., ZAGAME P., *Les nouvelles théories de la croissance en application: l'évaluation des politiques structurelles – Le cas du protocole de Kyoto*, Revue Économique, Vol. 56, No. 5, Paris, Septembre 2005.

Fraunhofer-ISI, *A comparison of thermal building regulations in the European Union*, MURE database case study No. 1, EC SAVE programme, 1999.

FREEMAN C., SOETE L., *The economics of industrial development*, MIT Press, 1997.

FRIEDRICH R. and BICKEL P. (eds.), *Environmental external costs of transport*, Springer-Verlag, Berlin, 2001.

GALBRAITH C.S., MERRILL G.B., *The politics of forecasting – Managing the truth*, California Management Review, Vol. 38, 1996.

GALLOPIN G., HAMMOND A., RASKIN P., SWART R., *Branch points: Global scenarios and human choice*, Stockholm Environment Institute, Scenario Group, SEI PoleStar Series Report No. 7, Sweden, 1997.

GALVIN R., *Science roadmaps*, Science, Vol. 280, No. 5365, 1998.

GAUDIN Th., *La prospective*, Que sais-je, PUF, Paris, 2005.

GAVIGAN J.P., SCAPOLO F., *Matching methods to the mission: A comparison of national foresight exercises*, Foresight, Vol. 1, No. 6, 1999.

GIOVANNINI B., BARANZANI A., *Energy modelling beyond economics and technology*, CUEPE, Genève, 1997.

GIRAUD P-N., Commissariat général du Plan, *Effet de serre, modélisation économique et décision politique*, Documentation française, Paris, 2002.

GIULIANI J-D., *Marchands d'influence*, Seuil, Paris, 1991.

GNESSOTO N., GREVI G., *The new global puzzle – What world for the EU in 2025?*, ISS, Paris, 2006.

GODET M., *Creating Futures – Scenario planning as a strategic management tool*, Economica, Paris, 2001.

GODET M., *The art of scenarios and strategic planning: Tools and pitfalls*, Technological Forecasting and Social Change, 2000.

GOLDEMBERG J., JOHANSSON T., REDDY A., WILLIAMS R., *Energy for a sustainable world*, Wiley, New York, 1988.

GORDON T.J., GLENN J.C., *Futures research methodology*, Millenium project of the American Council for the United Nations University, July, 2003.

GOULDER L.H., SCHNEIDER S.H., *Induced technological change and the attractiveness of CO_2 abatement policies*, Resource and energy economics, Elsevier, 1999.

GREENBERGER M., CRENSON M., CRISSEY B., *Models in the policy process – Public decision making in the computer area*, Russell Sage Foundation, New York, 1976.

GREENWOOD J., *Interest representation in the European Union*, Palgrave, 2007.

GRILICHES Z, *Issues in assessing the contribution of Research and Development to productivity growth*, Bell Journal of Economics, 1979.

GROSSMAN G.M., HELPMAN E., *Innovation and growth in the global economy*, Cambridge, MIT Press, 1991.

GRUBB M., EDMONDS J.E., *The cost of limiting fossil fuel CO_2 emissions – a survey and analysis*, Annual Review of Energy and Environment, Vol. 18, 1997.

GRUBLER A., NAKICENOVIC N., VICTOR D., *Dynamics of energy technologies and global change*, 1999, Energy Policy, Vol. 27, 1999.

GUILMOT J.F., McGLUE D., VALETTE P., WAETERLOOS C., *Energy 2000*, European Commission, Luxembourg, 1986.

HAMMITT J., *Analytic methods for environmental regulatiosn in the United States: promises, pitfalls and politics*, Int. J. Risk Assessment and Management, Inderscience, Vol. 1, No. 1/2, 2000.

Herry/Trafico, *Externe kosten in Güterverkehr in Österreich*, 2000.

HOHMEYER O., OTTINGER R.L., *Social costs of energy*, Springer-Verlag, Berlin, 1994.

HOHMEYER O., OTTINGER R.L., RENNINGS K., *Social costs and sustainability*, Springer-Verlag, Berlin, 1996.

HOURCADE J.C., JACCARD M., BATAILLE C., GHERSI F., *Hybrid modelling: New answers to old challenges*, The Energy Journal, Special issue, IAEE, Cleveland, 2006

Institute for Prospective Technological Studies (IPTS-JRC), *Preliminary analysis of the implementation of a EU-wide permit system on CO_2 Emissions abatement costs – Results from POLES model*, Study for the Commission (DG Environment), April 2000.

International Energy Agency, *Energy Technology Perspectives – Scenarios and Strategies to 2050*, OECD, Paris, 2006

International Energy Agency, *Energy to 2050 – Scenarios for a sustainable future*, OECD, Paris, 2003.

International Energy Agency, *Enhancing the market deployment of energy technology: a survey of eight technologies*, OECD, Paris, 1997.

International Energy Agency (C.O. WENE), *Experience curves for energy technology policy*, OECD, Paris, 2000.

International Energy Agency, *Experience curves: A tool for energy policy analysis and design*, Workshop, Paris, 22-24 January 2003.

International Energy Agency, *World Energy Outlook*, Paris, 2003 and 2006.

International Energy Agency, *World energy statistics*, Paris, 1999.

International Institute for Applied System Analysis, *Global energy perspectives*, Austria, 1998.

IPCC, *Climate Change – Third assessment report: Summary for policymakers*, WMO/UNEP/IPCC, 2001.

IPCC, *Climate Change – Fourth assessment report: Summary for policymakers*, WMO/UNEP/IPCC, 2007.

IPCC, *Special report on emissions scenarios*, Working Group III, Cambridge University Press, 2000.

ITALIANER A., D'ALCANTARA G., ZAGAME P., *The HERMES model for the Member States of the European Community*, in HERMES: Harmonised Econometric Research for Modelling Economic Systems, EC-North Holland, Luxembourg, 1993.

JAEGER C., RENN O., ROSA E., WEBLER T., *Risk, Uncertainty and rational action*, Earthscan, London, 2001.

JOERGES C., NEYER J., *From intergovernmental bargaining to deliberative political processes: the constitutionalisation of comitology*, European Law Journal, Vol. 3, No. 3, September 1997.

JORGENSON D., *Deconstructing the computer, measuring and sustaining the new economy*, The National Academies Press, Washington D.C., 2005.

KEMP R., *Technology and transition to environmental sustainability: the problem of technological regime shifts*, Futures, Vol. 26, 1994.

KEPPLER J.H., BOURBONNAIS R., GIROD J. (eds.), *The econometrics of energy systems*, Palgrave Macmillan, New York, 2007

KEYNES J.M., *General theory of employment, interest and money*, Cambridge University Press, 1936.

KLEIN L.R., *The making of national economic forecasts*, Edward Elger, London, 2009.

KLEINER A., *The man who saw the future*, Strategy and Business, Spring 2003.

KNIGHT F.H., *Risk, uncertainty, and profit*, Houghton Mifflin Company, Boston MA, 1921.

KOREISHA S., STOBAUGH R., *Limits to models*, in Energy Future, Report of the Energy project at the Harvard Business School, Random House, New York, 1979.

KOSTOFF R., SCHALLER R., *Science and technology roadmaps*, IEEE Transactions on Engineering Management, 2001.

KRIGE J., GUZZETTI L., *History of European S&T cooperation*, OPOCE, Brussels-Luxembourg, 1997.

KUHN Th., *The structure of scientific revolutions*, University of Chicago Press, 1962.

LAMFALUSSY A., BERNARD L., CABRAL A., *The Euro-zone: A new economic entity?*, Bruylant, Bruxelles, 1999.

LAPILLONNE B., CHATEAU B., *The MEDEE model for long term energy demand forecasting*, Socio-Economic Planning Sciences, Vol. 15, issue 2, 1981.

LAPILLONNE B., CHATEAU B., *The MEDEE models for long-term energy demand forecasting*, in Energy systems analysis, International Conference of Dublin, 9-11/10/1979, D. Reidel-Kluwer, ECSC-EEC-EAEC, Brussels and Luxemburg, 1980

LAPONCHE B., JAMET B., COLOMBIER M., ATTALI S., *Maîtrise de l'énergie pour un monde vivable*, ICE, Paris, 1997.

LATOUR B., *Science in action*, Harvard University Press, 1987.

LAWTON SMITH H., *The regulation of science and technology*, Basingstoke Palgrave Macmillan, 2001.

LESOURNE J., *Methodological problems in long-term energy modelling*, International Conference of Dublin *Energy systems analysis*, 9-11/10/1979, D. Reidel-Kluwer, ECSC-EEC-EAEC, Brussels and Luxemburg, 1980.

LESOURNE J., STOFFAES Ch., *La prospective stratégique d'entreprise – Concepts et études de cas*, Intersections, Paris, 1996.

LORD C., MAGNETTE P., *Notes towards a general theory of legitimacy in the European Union*, ESRC Working Paper 39/02, Sussex European Institute, 2001.

LORENZI J.-H., BOURLES J., *Le choc du progrès technique*, Economica, Paris, 1995.

LÖSCHEL A., *Technological change in economic models of environmental policy: a survey*, Ecological Economics No. 43, 2002.

LUCAS N., *Energy and the European Communities*, Europa Press, 1977.

MANNE A., *The use and role of models: Long-term energy projections for the USA*, International Conference of Dublin *Energy systems analysis*, 9-11/10/1979, D. Reidel-Kluwer, ECSC-EEC-EAEC, Brussels and Luxemburg, 1980.

MANNE A., RICHELS R., *Environmental modelling and assessment*, Springer, The Netherlands, 1997.

MANNE A., RICHELS R, WEYANT J., *Energy policy modelling: a survey*, Operations Research, Vol. 27, No. 1, 1979.

MARCHIPONT J.-F. and ROSSETTI di VALDALBERO D., *External costs: the European Commission's approach in the area of research*, Pollution atmosphérique, Paris, December 2001.

MARTIN B.R., IRVINE J., *Research foresight: Priority-setting in Science*, Pinter, London, 1989.

MEADOWS D.H., MEADOWS D., RANDERS J., *Beyond the limits – Global collapse or sustainable development*, Earthscan Publications, London, 1992

MEADOWS D.H., MEADOWS D., RANDERS J., BEHRENS W., *Limits to Growth*, Universe Books, New York, 1972.

MENY Y., MULLER P., QUERMONNE J-L., *Politiques publiques en Europe*, L'Harmattan, Paris, 1995

Ministère français de l'économie, des finances et de l'industrie, *Technologies clés 2010*, Paris, 2006.

MESAROVIC M., PESTEL E., *Mankind at the turning point*, Club of Rome, 1974.

MEUNIER S., KALYPSO N., *Who speaks for Europe? The delegation of trade authority in the EU*, Journal of Common Market Studies, Vol. 37, No. 3, September 1999.

MICHEL H., *Lobbyistes et lobbying de l'Union européenne*, Presses universitaires de Strasbourg, 2006.

MINTZBERG H., *The rise and fall of strategic planning*, The Free Press, New York, 1994.

MOLINA M.J., ROWLAND R.F., *Stratospheric ozone sink for chlorofluoromethanes: chlorine atom catalyzed destruction of ozone*, Nature, Vol. 249, 1974.

MORGENSTERN R., *Economic analysis at EPA: Assessing regulatory impacts*, Resources for the Future, Washington, DC, 1997.

MORIN E., *An I de l'ère écologique*, Nouvel Observateur, Paris, 1972.

MURE model estimation on energy efficiency in the EC, SAVE programme, Study for the Commission (DG XVII-Energy), March 1998.

NAKICENOVIC N., NORDHAUS W.D., RICHELS R., TOTH F.L., *Integrative assessment of mitigation, impacts, and adaptation to climate change*, International Institute for Applied Systems Analysis (IIASA), Laxenburg-Austria, 1994.

NAKICENOVIC N., GRUBLER A., MAC DONALD A., *Global energy perspectives*, Cambridge University press, 1998.

National Technical University of Athens, *A business as usual energy scenario for pre-Kyoto*, Study for the Commission (DG XVII–Energy), April 1997.

National Technical University of Athens, *The economic effects of EU-wide industry-level emission trading to reduce greenhouse gases – Results from PRIMES model*, Study for the Commission (DG Environment), May 2000.

NELSON R., WINTER S., *An evolutionary theory of economic change*, Belknap Press and Harvard Press, London and Massachusetts, 1982.

NEUREITER N., *Challenges of providing scientific advices to governments*, SRC, Ispra, 26/10/2009.

NORDHAUS W., *Critical assumptions in the Stern Review on climate change*, Science, Vol. 317, 13 July 2007.

NORTH D.C., *Institutions, institutional change and economic performance*, Cambridge University Press, UK, 1990.

OECD, *Frascati Manual*, Paris, 2002.

OECD, *Improving governance for sustainable development*, Seminar, Paris, 22-23/11/2001.

OECD, *Is there a new economy*, Paris, 2000.

OECD, *Main science and technology indicators*, Paris, 2006.

OWENS S. and RAYNER T., *When knowledge matters: the role and influence of the Royal Commission on Environmental Pollution*, Journal of Environmental Policy and Planning, 1, 1999.

PEARCE D., *Cost-benefit analysis and environmental policy*, Oxford Review of Economic Policy, Vol. 14, No. 4, winter 1998.

PEARCE D., MARKANDYA A., BARBIER E., *Blueprint for a green economy*, Earthscan Publications, London, 1989.

PEARCE D., MARKANDYA A., *The benefits of environmental policy*, OECD, Paris, 1989.

PEARCE D., *The role of carbon taxes in adjusting to global warming*, The Economic Journal, Vol. 101, 1991.

PETIT P., SOETE L., *Technology and the future of European employment*, Edward Elgar Publishing Ltd, Cheltenham, 2000.

PILAVACHI P.A., DALAMAGA Th., ROSSETTI di VALDALBERO D., GUILMOT J.-F., *Ex-post evaluation of European energy models*, Energy Policy, Vol. 36, 2008.

PORTER A., *Technology futures analysis: Toward integration of the field and new methods*, 2004.

RASKIN P., BANURI T., GALLOPIN G., GUTMAN P., HAMMOND A., KATES R., STEWART R., *Great transition: The promise and lure of the times ahead*, Stockholm Environment Institute, Boston, 2002.

RASKIN P., GALLOPIN G., GUTMAN P., HAMMOND A., SWART R., *Bending the curve: Towards global sustainability*, Stockholm Environment Institute and Global Scenario Group, SEI PoleStar Series Report No. 8, Stockholm, 1998.

RHINARD M., *The democratic legitimacy of the European Union committee system*, Governance, Vol. 15, April 2002.

RICHARDSON J., *European Union, power and policy-making*, Routledge, 2006.

RINGLAND G., *Scenarios in business*, John Wiley & Sons, Chichester, 2002

ROQUEPLO Ph., *Penser la technique*, Seuil, Paris, 1983.

ROQUEPLO Ph., *Climats sous surveillance – Limites et conditions de l'expert*, Paris, 1993.

ROMER P.M., *Endogenous technological change*, Journal of Political Economy, No. 98, 1990.

ROSENBERG N., *Perspective on technology*, Cambridge University Press, 1976.

ROSENBERG N., *Inside the black box: technology and economics*, Cambridge University Press, 1982.

ROSSETTI di VALDALBERO D., *Les nouveaux paradigmes énergétiques et environnementaux: la place de l'Europe*, Revue de l'Énergie, No. 584, Paris, juillet-août 2008.

ROSSETTI di VALDALBERO D., BAER J.M., *Les coûts sociaux en Europe – Recherche, énergie, transport et environnement*, IEPF, No. 74, Québec, 1er trimestre 2007.

ROSSETTI di VALDALBERO D., BALEN J., *Energy and transport: pricing externalities in the European Union, Setting concepts into motion: sustainable development and R&D policies*, OSTC, Proceedings, Brussels, 2001.

ROSSETTI di VALDALBERO D., *Energy market regulation: Impacts of EU research*, in BÖHRINGER Ch., LANGE A., *Applied research in environmental economics*, ZEW-Physica-Verlag, Heidelberg, 2005.

ROSSETTI di VALDALBERO D., *Energy technologies and climate change: a world and European outlook*, Global Foundation, Kluwer, New York, 2001.

ROSSETTI di VALDALBERO D., VALETTE P., *Environmental and health costs in the European Union: Research and policy*, Encyclopedia of environmental health, Elsevier, 2009.

ROSSETTI di VALDALBERO D., *L'Europe et le changement climatique: Actualité et prospective*, Revue de l'Énergie, No. 534, Paris, février 2002.

ROSSETTI di VALDALBERO D., *Modelling in energy economics and policy*, 6th European Conference of the International Association of Energy Economics, Zurich, 2/9/2004.

ROSSETTI di VALDALBERO D., *Quantification and tentative internalisation of energy external costs in the EU*, in SPITZER C., SCHMOCKER U., DANG V.N. (ed.), *Probabilistic Safety Assessment and Management*, Vol. 4, Springer, London, 2004.

ROSSETTI di VALDALBERO D., SCHMITZ B., RALDOW W., POIREAU M., *European Union energy research*, Revue de l'Énergie, No. 576, Paris, Mars-Avril 2007.

ROUSSEAU J-J, *Discours sur les sciences et les arts*, Paris, 1750.

RUBIN R., ROMBERG E., *Energy models for the European Community*, in A. STRUB (ed.), *Energy models for the European Community*, IPC Science and Technology Press, 1979.

SENDEN L., *Soft law in European Community law*, Hart Publishing, UK, 1994

Shell, *Energy needs, choice and possibilities: scenarios to 2050*, London, 2001.

Shell, *Global scenarios to 2025 – The future business environment: trends, trade-offs, and choices*, London 2005.

SALO A., GUSTAFSSON R., KAAKKOLAMMI I., GUSTAFSSON T., *Participatory development of future-oriented policy configurations for climate change and mobile gaming*, 2004.

SALOMON J.-J., *Le destin technologique*, Gallimard, Paris, 1992.

SAY J.-B., *Traité d'économie politique*, Paris, 1803.

SCHADE B., WIESENTHAL T., *Comparison of long-term world energy studies – assumptions and results from four world energy models*, IPTS Working Paper, Sevilla, March 2007.

SCHARIOTH J., HUBER M., SCHULZ K., PALLAS M., *Horizons 2020 – A thought-provoking look at the future*, study commissioned by Siemens, TNS Infratest, Munich, 2005.

SCHARPF F., *Notes toward a theory of multilevel governing in Europe*, Max-Planck-Institut für Gesellschaftforschung, Discussion Paper, 5/2000.

SCHRAGE M., *Science must be more political*, Financial Times, 26/9/2007.

SCHUMPETER J., *Business cycles: a theoretical, historical and statistical analysis of the capitalist process*, Mc Graw Hill Book Company, New York, 1939.

SCHURR S., NETSCHERT B., *Energy in the American economy: 1859-1975*, John Hopkins Press, Baltimore, 1960.

SCHWARTZ P., *La planification stratégique par scénarios*, Futuribles, Paris, Mai 1993.

SCHWARTZ P., *The art of the long view: Planning for the future in an uncertain world*, Doubleday, New York, 1991.

SLAUGHTER R.A., *Future studies as an intellectual and applied discipline*, American Behavioural Scientist, Vol. 42, No. 3, 1998.

SLAUGHTER R.A., *The foresight principle: Cultural recovery in the 21st Century*, Adamantine Press, London, 1995.

SPENCE D., *The European Commission*, John Harper, UK, 2006.

SPENCER W., WILSON L., DOERING R., *The semiconductor technology roadmap*, Future Fab International 18, USA, 12 January 2005.

SRUENCK Ch., *Why is there no mad cow disease in the United States? Comparing the politics of food safety in Europe and the U.S.*, Working Paper, Institute of European Studies, University of California, Berkeley, December 2001.

STASTNY D., *Economics and politics: Mutual relationship*, Working Paper, Austrian Scholars Conference, 15-16 March 2002.

STERN N., *Review of economics of climate change (Stern Review)*, UK Chancellor of the Exchequer, London, 2007.

STEUNENBERG B., KOBOLDT B., SCHMIDTCHEN D., *Comitology and the balance of power in the European Union*, International Review of Law and Economics, 1996.

STOBAUGH R., YERGIN D. (eds.), *Energy future*, Random House, New York, 1979.

STRUB A., *Energy systems analysis*, International Conference of Dublin, 9-11/10/1979, D. Reidel-Kluwer, ECSC-EEC-EAEC, Brussels and Luxemburg, 1980.

TERES and TERES II, The European Renewable Energy Study, The prospects for renewable energy in 30 European Countries from 1995-2020, ALTENER programme, Study for the Commission (DG XVII-Energy), November 1996.

TOSATO G., *Applied systems analyses, technology and research, energy models*, Rome, March 2007.

THALMANN Ph., VIGUIER L., *Technology in climate policy and climate models*, Energy Policy, Special edition, Vol. 35, No. 11, November 2007.

TOFFLER A., TOFFLER H., *The third wave*, Bantam books, New York, 1980.

UK House of Commons, *Governing the Future*, second report of session 2006-2007, London, 2007.

UNIDO, *Technology foresight manual*, Vienna, 2005.

United Nations Foundation, Expert group on energy efficiency, *Realizing the potential of energy efficiency*, Washington, 2007.

TORGERSEN H., *The real and perceived risks of genetically modified organisms*, EMBO, 2004.

US Department of Energy (EIA), *International Energy Outlook*, Washington, 2001 and 2006.

VALETTE P., European Commission, DG Research, *EU 2025*, internal note, 21/11/2007.

VAN DER HEIJDEN K., BRADFIELD R., BURT G., CAIRNS G., *The Sixth Sense*, Wiley, New York, 2002.

VAN DER HEIJDEN K., SCHUTTE P., *Key questions for designing scenario applications*, Scenario and Strategy Planning, Vol. 1, No. 6, 2002.

VAN DER VOORT E., DONNI E., THONET C., BOIS D'ENGHIEN E., DECHAMPS C., GUILMOT J.-F., *Energy supply – Modelling package EFOM-12C Mark I*, Babay, Louvain-la-Neuve, 1984

VAN DER WEE H., *Histoire économique mondiale*, Academia-Bruylant, Bruxelles, 1990.

VAN GOETHEM G. *et al.*, *Euratom research and training in reactor systems and fuel cycles: EU response to some of the main S/T challenges in nuclear fission*, FISA-2006 Conference *EU Research and Training in Reactor Systems*, Luxembourg, 13-16 March 2006.

VOS H., WENDLER F., *Food safety regulation in Europe*, Intersentia, Antwerpen-Oxford, 2006.

VTT, *Energy visions 2030 for Finland*, Helsinki, 2001.

WACK P., *Scenarios: Uncharted waters ahead*, Harvard Business Review, September-October. 1985.

WACK P., *La planification par scénarios*, Futuribles No. 99, Paris, mai 1986.

WALLACE H., WALLACE W., POLLACK M.A., *Policy-making in the European Union*, Oxford University Press, 2005.

WALRAS M.E.L., *Éléments d'économie politique pure*, Lausanne, 1874.

WAN H.Y., *Causally indeterminate models via multi-valued differential equations, Bifurcation theory and applications in scientific disciplines*, Annals of New York Academy of Sciences, Vol. 316, 1979.

WEBSTER A., *Foresight as a tool for the management of knowledge flows*, Report for EC STRATA workshop, Brussels, 22-23/4/2002.

WEILER J., *Epilogue: Comitology as revolution – Infranationalism, constitutionalism and democracy*, in JOERGES C., VOS E. (eds.), *EU Committees: Social regulation, law and politics*, Hart Publishing, Oxford, 1999.

WEILER J., *Problems of legitimacy in post 1992 Europe*, Aussenwirtschaft 46, 1991.

WEINBERG A., *Limits to energy modelling*, International Conference of Dublin *Energy systems analysis*, 9-11/10/1979, D. Reidel-Kluwer, ECSC-EEC-EAEC, Brussels and Luxemburg, 1980.

WELSCH H., *Recycling of carbon/energy taxes and the labour market*, Environmental and resource economics, Vol. 8, 1996.

WESSELS W., *Comitology: fusion in action – Politico-administrative trends in the European Union system*, Journal of European Public Policy, Vol. 5, No. 2, 1998.

WESTLAKE M., GALLOWAY D., *The Council of the European Union*, John Harper, 2006.

WEYANT J., *The role of models in the oil and gas pricing debates*, Energy Modelling Forum, Stanford University, April 1979.

WHITELEY M., BESS M., *Renewable energy technologies and the European industry*, International Journal of Global Energy Issues, 14, 2000.

WING I.S, ECKAUS R.S., *The implications of the historical decline in US energy intensity for long-run CO_2 emission projections*, Energy Policy, Special edition, Ph. THALMANN and L. VIGUIER (ed.), *Technology in climate policy and climate models*, Vol. 35, No. 11, November 2007.

World Energy Council (WEC)/International Institute for Applied for Systems Analysis (IIASA), *Global energy perspectives*, IIASA, Laxenburg, 1998.

ZAGAME P., *L'expérience française de modélisation macro-économétrique: bilan et perspectives*, Revue d'économie politique, No. 5, Paris, 1987.